CONTINUOUS SOFTWARE ENGINEERING

CONTINUOUS SOFTWARE ENGINEERING

Edited by:

Ivan Stanimirović

ARCLER PRESS

www.arclerpress.com

Continuous Software Engineering

Ivan Stanimirović

Arcler Press
2010 Winston Park Drive,
2nd Floor
Oakville, ON L6H 5R7
Canada
www.arclerpress.com
Tel: 001-289-291-7705
001-905-616-2116
Fax: 001-289-291-7601
Email: orders@arclereducation.com

ISBN: 978-1-77361-069-6 (Hardcover)

Library of Congress Cataloging-in-Publication Data

A CIP record for this book has been applied for and is available from Library and Archives Canada.

DECLARATION

Some content or chapters in this book are open access copyright free published research work, which is published under Creative Commons License and are indicated with the citation. We are thankful to the publishers and authors of the content and chapters as without them this book wouldn't have been possible.

ABOUT THE EDITOR

Ivan Stanimirovic gained his PhD from University of Niš, Serbia in 2013. His work spans from multi-objective optimization methods to applications of generalized matrix inverses in areas such as image processing and computer graphics and visualisations. He is currently working as an Assistant professor at Faculty of Sciences and Mathematics at University of Niš on computing generalized matrix inverses and its applications.

TABLE OF CONTENTS

List of Contributors xiii
Preface xv

Chapter 1 Analysis of Integration Mechanisms of Tools for Continuous Software Engineering 1
1. Introduction 1
2. Current State of the Art 3
3. Design of The Process 14
Conclusions 23

Chapter 2 Automation of Web Compatibility Tests in a Continuous Software Development Environment 25
1. Web Compatibility Tests 25
2. Proposed Technique 27
3. Results Obtained 30
Conclusion 30

Chapter 3 Continuous Software Delivery in a Cloud Service 33
1. Introduction 34
2. Methodology of Software Development 37
3. Analysis of Agile Development Frameworks 47
4. Configuration of Scrum 54

Chapter 4 Technological Background of Continuous Delivery in Cloud Computing 57
1. Cloud Computing 57
2. Previous Concepts 60
3. Tools For Continuous Delivery 64

Chapter 5 **Design and Implementation of Continuous Delivery in Cloud Computing** **67**

1. Continuous Delivery 67
2. Development Pipeline 68
3. Implementation Of Continuous Delivery 71
4. Implementation 76

Chapter 6 **Validation of Continuous Delivery in Cloud Computing in the Real Environment** **77**

1. Introduction To The Environment 77
2. Validation 78
3. Development Environment 81
4. Variations On Initial Planning 82

Chapter 7 **Adaptive Development Processes For Agile Software Engineering: Methodology And Practice** **83**

1. Introduction 83
2. Context Of Agile Projects 85
3. Qualdev Process 88
4. Results And Xp Methodology Suggestions 93
5. Conclusions and Future Work 105

Chapter 8 **Conclusions, Enhancements and Extensions** **107**

Chapter 9 **Hora: Architecture-Aware Online Failure Prediction** **111**

Abstract 111

1. Introduction 112
2. Motivating Example 114
3. The Hora Prediction Approach 116
4. Evaluation 127
5. Evaluation Metrics 130
6. Related Work 144

Conclusion 147

Acknowledgment 148

References 149

Chapter 10 Continuous Delivery: Overcoming Adoption Challenges 157

Abstract.. 157

1. Introduction 158
2. The Context... 160
3. Strategies To Overcome Cd Adoption Challenges 165
4. Challenges For Research 184
5. Related Work 190
6. Summary... 193

Acknowledgments.. 195

References 196

Chapter 11 How Do Software Development Teams Manage Technical Debt? – An Empirical Study 201

Abstract.. 201

1. Introduction 202
2. Background... 205
3. Research Process... 211
4. Results 222
5. Summary of The Cases 239
6. Technical Debt Management Framework 245
7. Discussion 251

Conclusion.. 264

Acknowledgments.. 264

References 270

Chapter 12 Challenges And Success Factors For Large-Scale Agile Transformations: A Systematic Literature Review 279

Abstract.. 279

1. Introduction 280
2. Background... 281
3. Research Method 284
4. Results 294
5. Discussion 333
6. Conclusions 342

Acknowledgment... 343

References 343

References 349

Citations 353

Index 355

LIST OF CONTRIBUTORS

Ivan Stanimirović

Teerat Pitakrat

University of Stuttgart, Institute of Software Technology, Reliable Software Systems, Germany

Dušan Okanović

University of Stuttgart, Institute of Software Technology, Reliable Software Systems, Germany

André van Hoorn

University of Stuttgart, Institute of Software Technology, Reliable Software Systems, Germany

Lars Grunske

Humboldt University Berlin, Department of Computer Science, Software Engineering, Germany

Lianping Chen

Lianping Chen Limited, Dublin, Ireland

Jesse Yli-Huumo

Lappeenranta University of Technology, School of Business and Management, Department of Innovation and Software, PO Box 20, Skinnarilankatu 34, Lappeenranta FI-53851, Finland

Andrey Maglyas

Lappeenranta University of Technology, School of Business and Management, Department of Innovation and Software, PO Box 20, Skinnarilankatu 34, Lappeenranta FI-53851, Finland

Kari Smolander

Aalto University, School of Science, Department of Computer Science, P.O.Box 15400, FI-00076 Aalto, Finland

Kim Dikert

School of Science, Department of Computer Science and Engineering, Aalto University, Finland

Maria Paasivaaraa

Massachusetts Institute of Technology, Sloan School of Management, United States

Casper Lassenius

School of Science, Department of Computer Science and Engineering, Aalto University, Finland

Massachusetts Institute of Technology, Sloan School of Management, United States

PREFACE

At present, most of the software projects that are developed, are run using methodologies and procedures, which causes the team in charge of the project end up launching products and services with failures. This is why companies need to consider the quality of software developed from a perspective that covers not only the final quality of the product, but a more general perspective which covers the entire development process. It is necessary to make use of methodologies, procedures and standards to be able to develop and launch a software product of quality, with sufficient certainty to obtain a lower final cost and risk.

The typical software life cycles have processes that isolate the tasks to be carried out by the different teams of the project. These are the development, testing, product (business) administration and operations, so that there is an important barrier between these.

In addition, during the software life cycle, it passes through different environments (the development, testing, production, etc.), so that it is a source of errors that are common to produce software in an environment making the software dependent on it and at the time of in the environment in which the users are going to interact with the application or service, this it does not work correctly. One of the intermediate steps between these two environments is the tests, where the correct functioning of the software will be verified, that is why it is it is advisable to carry them out in an environment as close as possible to that of production to mitigate the possible errors produced by the difference of scenarios.

In addition to this scenario difference problem, this process of bringing the software to the production environment, commonly referred to as production start-up, is completely manual, of errors, unreliable and difficult to repeat.

All these disadvantages cause that this process that goes from the software development until its production can lead to take in most cases weeks, delay that for some companies can translate into a high cost of opportunities, or what is the same a high economic cost.

To overcome all these problems, new methodologies and software development:

- DevOps defined as a new working philosophy in which barriers are broken previously discussed by the different software project teams, through the creation of an ecosystem that favors the communication, collaboration and integration of all these equipment.
- Continuous delivery or continuous supply [1] as a form of software development that makes use of good practices for the software to be built so that it can be put into production in an agile and less risky way, carrying out frequent launches that allow the cycle between the development of the software and the supply of this to the user is as short as possible, allowing even daily deployments to be carried out faster feedback from users. Throughout the text will be used indistinctly, as being synonymous, the terms continuous delivery and continuous supply.

One way to facilitate the implementation of the continuous delivery model is to integrate of an automated, repeatable and reliable process that encompasses the set of common tasks involved in the deployment and development of software. Tasks ranging from the developer uploads to the software repository of the project until it is put into production, producing a rapid feedback, so that the developments that are not valid for production be discarded as quickly as possible.

In addition, this automation allows to solve in part the barrier previously commented between the different project teams, reducing the delay between phases that depend on different groups, making the process tasks carried out by these groups can be performed with the press of a button.

Continuous software engineering is an approach in which equipment maintains the production of software in short cycles of time, ensuring that the product can be released reliably in any moment. Today, this approach is increasingly being used in organizations, especially those that develop web or mobile. However, by releasing versions of the product with greater frequency, more defects emerge in the same. This is mainly due to that the time to perform the test cycles are very short. One of the challenges that currently exists is the acceleration of testing on the interface such as web compatibility. A technique used in a large software development company is presented to automate web compatibility testing, automation image comparison when doing cross-tests between different browsers. The results indicate that the proposed technique adapts to the requirements of the processes of continuous development, increasing the performance and speed of this type of tests.

Editor

CHAPTER 1

ANALYSIS OF INTEGRATION MECHANISMS OF TOOLS FOR CONTINUOUS SOFTWARE ENGINEERING

Ivan Stanimirović

INTRODUCTION

The Continuous Software Development (CSD), is an approach of the Engineering of Software, in which, the equipment produces software in very short cycles of time, ensuring the release of different versions of the product at any time [1]. Within this field arise different techniques and methodologies, for example: Continuous integration, continuous deployment, continuous delivery and continuous testing.

The CSD is a key factor in organizations as a differentiator of competition, which also enables improvements to be achieved quickly and efficiently in deliveries to customers, providing them with a reliable product [2].

However, one of the main challenges is to maintain high product quality software. When the system is deployed more frequently, more defects in the product [3], [4]. This is mainly due to the fact that time to perform the test cycles are very short.

As a solution to this problem researchers have tried to maintain the quality of the product software using automated testing. By means of the same, seeks to shorten delivery times of products [5]. Thus, for example, Google generates lots of automated tests using specialized tools [6], acquiring a greater coverage of tests before releasing a new version of the product. By contrast, Mike Cohn and Anand Bagmar, put the evidence units and those of acceptance as a basis in the test pyramids [7] [8]; and the graphical interface (UI) tests are in the last place

This leads many organizations to put their best efforts into automating unit tests and spatially, acceptance tests. Thus, for example, that small software development teams are tests of manual acceptance [9].

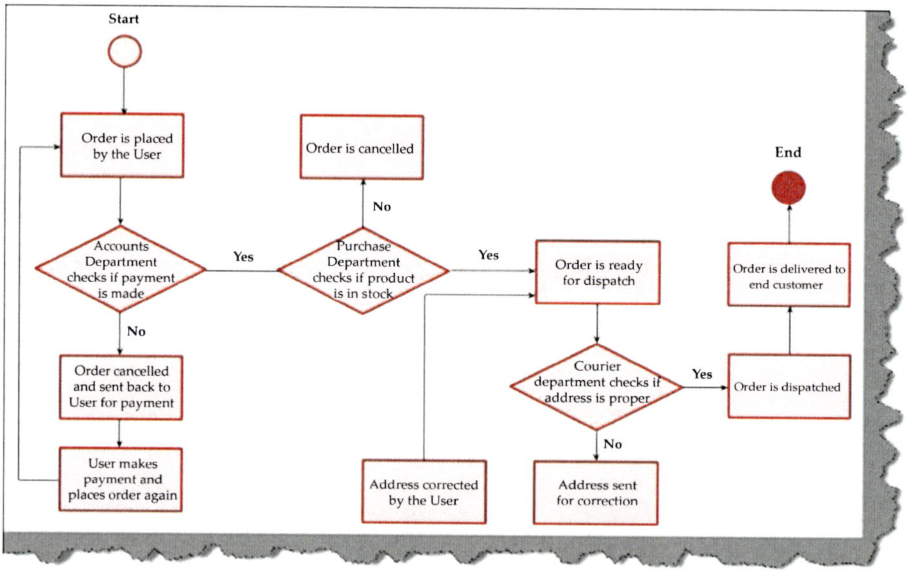

However, for other companies, it is vital that the product under development be compatible in different browsers both on a computer and in a mobile device. The tests used to determine the compatibility of an application in different browsers are called "compatibility tests Web "or" cross-browser testing "(better known as CrossBrowser Testing). In this book researchers present a technique to automate the tests based on the comparison of images using the digital image processing.

The main objective of this chapter is to demonstrate that complete automation of the life cycle of an application by carrying out the integration of existing tools, which will be presented in the next chapter, so that allows the implementation of a continuous delivery model. The design and assembly of each one of the tasks will be detailed later that allow carrying a project from its new development to the production environment, feedback to all team members involved in the creation process software.

CURRENT STATE OF THE ART

In this section, researchers will carry out, on the one hand, the description of the continuous delivery process and the set of concepts that must be applied to achieve its implementation, and which will be implemented during the work, and on the other hand, it will focus on the analysis and description of existing tools to be used. This part is very important since it is a question of solving a process problem, the tools are the ones that support the automation of this, and therefore if there are tools for all the process and integrate, it will have managed to automate the whole process. It should be noted that the choice of open source tools.

The continuous delivery process

As described in the introduction, continuous delivery consists of software using good practices so that the software is built in a way that can be put into production in an agile and less risky way, being able to carry out frequent releases that allow the cycle between software development and supply to the user as short as possible, even allowing daily, which makes it possible to obtain feedback more quickly from the users.

Some of the practices that are carried out to achieve a continuous delivery process are the continuous execution of tests, continuous integration and above all the automation of all these practices.

Continuous integration

Continuous integration is considered a software development practice consisting of integrating the code generated by the different developers of a project on a frequent basis, being thus performed from one to several daily integrations.

In this way, an improvement of productivity and quality is provided in all software projects involved, allowing all code-producing equipment to receive the fastest possible way, being able to solve immediately the possible problems of integration introduced after the last change on the project repository. Therefore, an automated and repetitive process must be carried out in order to download the version control project, compile it, run the necessary tests to collect the errors and generate reports. All this process is usually carried out periodically or when a developer makes a change about the project that may be a source of errors.

Automation of tests

Whenever a software development is carried out, to ensure that a certain level of quality of the same, it is necessary to verify the set of requirements and functionalities that must provide this. This verification is carried out by the execution of some tests that try to find errors in the software and test its functionality, for everything this software testing is one of the most critical stages in the life cycle software.

As shown in the diagram below, the typical testing process encompasses the design, execution and analysis of results.

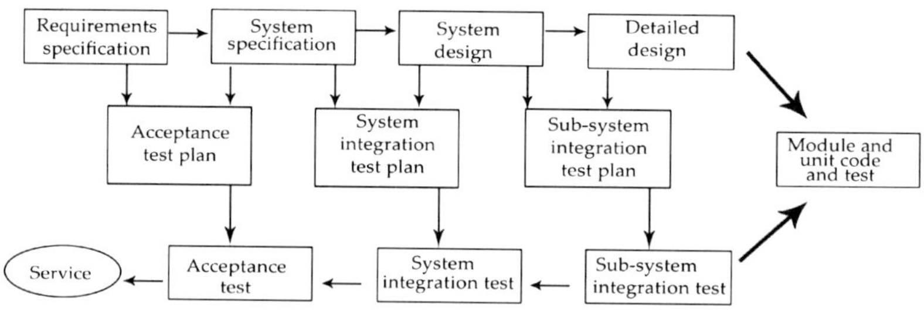

Figure 2: Phases of a test process

Automating these tests consists of using a set of test software that facilitates the automation of the last two phases of this process, the execution of the designed tests and the analysis of the results by comparing the results of the implementation with the expected results. The benefits of automating these tests can be summarized in:

- Obtain more reliable data due to the establishment of a repeatable process where errors of variation of the execution of the tests due to being carried by hand are avoided.
- Increased number of tests performed, reducing the time required perform them manually.
- Increased productivity by the testing team, since the time savings due to the automation of the execution phases and analysis of results is invested in the design phase of new tests.
- Possibility of carrying out more complicated tests, since some tests are too complex to be carried out manually. Validating like this characteristics of the systems that could not be otherwise validated.
- More stable code, because new code developments are carried out on versions that have already been tested, automated testing is simpler execute them frequently on the code.

Although the integration of these last two practices, in an automated process and repeatable, allows you to bring a project closer to a continuous delivery practice, this is not enough. Continuous delivery is also a change in how the team in charge of producing software thinks on the form of work. Instead of thinking that software should be over very long periods of time, fixing during this time things that make the system does not work, etc. You should think about carrying out a more agile process where the team smaller code changes more frequently, so you can solve smaller problems more quickly, all equipment must take a step more and think that the responsibility of getting a quality code is everyone's, so development, testing and putting into production must be a single task integrated into that.

Finally, once all the concepts described above have been assimilated, what is intended in this study is the design and implementation of an automated process that provides a strong loop and automated execution of certain phases of the development of an application to facilitate the ultimate goal of continuous delivery process. Thus, researchers want to achieve the automation of the following phases within the software development process.

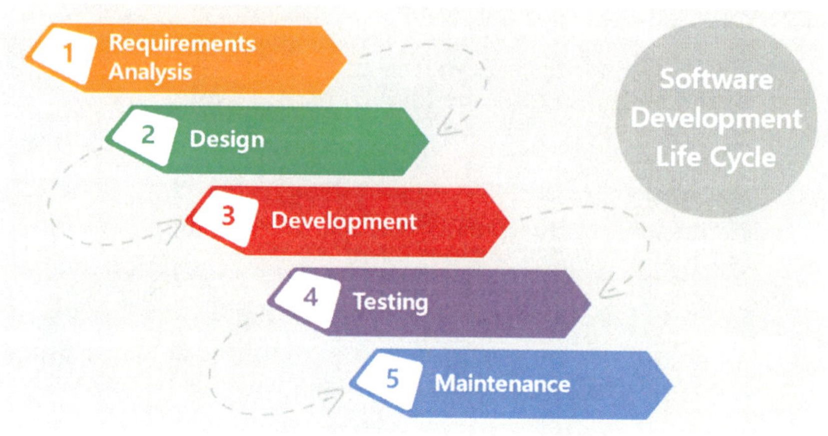

Figure 3: Phases of a software development process

One way to approach the implementation of this continuous delivery model is to rely on an abstraction called the Deployment Pipeline.

Deployment Pipeline tries to model the set of phases and automated tasks that are due In the process from obtaining version control software to that the software is in a state ready to be delivered to users, in default that its production is confirmed. The entry into the pipeline is a particular version control that leads to a construction of the project, which must pass through a sequence of tests that demonstrate the feasibility or otherwise of the project for its production. The main objective is to eliminate in production and get error feedback so that it can be solved as fast as possible.

This process begins when the team in charge of the project carries out an update in the version control system, at which point the construction and execution phase is triggered of unit tests. Once finished, this phase can return two results:

1. There is a failure in the unit tests and the project is not built, or the tests are correct but there is a failure during the construction of the project, notifying to interested parties that it has failed.
2. The tests are correct and the project is built, notifying the interested parties that the phase has been successfully completed and the execution of the next phase is triggered.

The next phase will be the execution of automated acceptance tests, again this phase may return two results:

1. A failure occurs in the acceptance tests, notifying the results to the concerned parties.
2. All acceptance tests are successful, notifying interested parties the results of the tests and that it is possible to move on to the next stage. In this case the phase will not be triggered automatically, but it will be necessary for the equipment to the good to move on to the next phase. The next phase consists of acceptance tests of the user, tests that are executed manually and that result in a report that provides the feedback to the team in charge of the software delivery.

Finally, the team, in view of all the results obtained, will evaluate if the software is ready to be put into production, in which case the final deployment of the application will be made.

This diagram summarizes the automated process that is to be integrated in this work and that allows an approximation to the implementation of a continuous delivery model. The design and integration of this process will occupy the following chapters of the book.

Finally, it is going to describe a concept that is common to confuse with the concept of delivery continuous, the concept of continuous deployment.

Both concepts can be implemented through a process formed by the same phases and following the same principles, but while continuous delivery has as its ultimate goal leave the software in a state ready for production, so that the equipment manager analyzes the results of all phases and decides by the push of a button deployment in the production environment, continuous deployment goes one step further.

Further automating this last step, so that its ultimate goal is to achieve carry out frequent deployments, being able to deploy applications with each change of the software project. As can be seen it is formed by:

- Sensors: Responsible for gathering information about the context in which the system.
- Inputs: Through which to enter the initial information to define the behavior.

- Actuators: Those responsible for carrying out the actions to adapt the system to the changes.
- Intent: Set of objectives, ie, the service it proposes to provide, and policies, set of rules that define the basic behavior context, which direct the actions to be taken by the system.
- Knowledge of the operation: Information that defines how the system should operate without intervention, such as configuration, mode of interpretation of the data provided by sensors, etc.
- Logic: part of the decision-making system based on the definition of the intention and the information collected by the sensors.

On the other hand, the functions that can have an Autonomous System are:

- Self-configuration: It is able to dynamically adapt to changes in the environment, making use of the policies provided by the system administrator. It can include deployment of new components, replication of these, changes of characteristics of operation, etc. Thus ensuring an increase in system productivity.
- Self-optimization: It is able to monitor and regulate system resources to satisfy the needs of the user.
- Self-regeneration: Is able to discover, diagnose and react to bad functioning of the system, taking corrective measures without affecting the system environment.
- Self-protection: Is able to identify and protect against hostile behaviors, such as denial of service attacks, unauthorized infiltration, etc. And take action corrective measures that allow the system to be less vulnerable.

Using these functions, Apache Brooklyn allows modeling applications as a set of micro-services, through entities grouped in the form of a hierarchical tree, and manage their running at run time.

It has been decided to use Apache Brooklyn in the implementation of the process described as being the only implementation of this type of tools.

Software Test Automation Tools

Junit

JUnit [4] is a familiar unit testing framework for Java that was developed by Erich Gamma and Kent Beck, and allows testing the execution of Java classes to see if the methods developed works as expected. It is free code software released under license Common Public version 1.0. Annotations were included starting with version 4 of JUnit, which are a set of keywords, placed in front of the methods, that allow to indicate to Junit specific instructions.

This tool has been chosen to carry out the unit tests of the project to process in the automated process. It has been chosen as the unit testing tool of Java used, which can easily be integrated with most environments and tools existing ones.

Selenium IDE

Selenium 2.0 [5] is a free code framework that allows the automation of tests on a website. This framework provides a recording / playback tool, carry out a web browsing session, generate a script from this navigation and subsequently run as a test to perform so-called functional tests of automated way.

This tool has been chosen to carry out the functional tests of the project to process. It has been chosen because it is relatively simple to use and is compatible with many languages, in particular with the Java language, besides allowing to export the tests to Java classes to be executed as if they were JUnit tests.

Apache JMeter

Apache JMeter [6] is a tool developed by The Apache Software Foundation, whose first version dates back to March 2001. It was designed as a tool to carry out performance testing and functional testing on web applications. At first it was thought as a tool for running stress tests on simple web applications

But its architecture has been developed to carry out more complex tests on databases, Perl script, servlets, FTP servers, etc.

This tool has been chosen to carry out the load tests in the process that is going to be integrated as an ideal and simple tool to carry out the load tests thanks to its architecture, and to generate colorful HTML reports that can be integrated into the view of the server that is going to be used.

Project Building Tools

Maven

Apache Maven [7] is one of the most used tools in the management and construction of projects Java software created in 2002 by Jason van Zyl, whose functionality resembles that of Apache Ant, PEAR, CPAN, etc.

This tool is based on the structure of the software project to be managed and a file describing all the properties of the project called POM (Project Object Model).

The latter file is written in XML format and contains information such as an identifier unique project, license, project members, project dependencies, repositories Maven artifacts (from which you can obtain the dependencies required by the project), plugins that allow to add functionality to the project and many others.

Maven defines a strict life cycle for an application so that execution of one of the phases of this cycle involves the execution of all the previous phases. In an orderly manner, the phases that define the life cycle of the project are:

- Compile: Compile the .java sources by generating the respective .class files
- Test: Execute existing JUnit test, if any test fails, do not continue with the next phase.
- Package: Creates a package with the files compiled in the format specified in the POM file.
- Install: Copy the generated package to a local Maven directory so that it can be used again in another project.
- Deploy: Copy the generated package to a remote server so that it can be used from new for any other project.

This project-building tool has been chosen for the standardization of the life, its dependency management system based on repositories with a large number of tools and frameworks, which will also allow you to run tests in an automated way.

Software Process Automation Tools

In this section, two similar tools will be presented and automation of the software process. After presenting them, it will conclude with a comparison of these with the purpose of opting for one of the two.

Go Continuous Delivery

Go CD (Continuous Delivery) [8] is an open source server written in Java, which aims to carry out a continuous delivery process, in other words a process of continuous integration along with a release management system, based on an Agent - Server. It was created by the ThroughtWorks group and it was not until February 25, 2014 when was released under Apache 2.0 license. You can process projects written in all types of languages of programming, being quite solvent with projects developed in Java.

The Agent - Server model allows the distribution of work over agents. This distribution of jobs is carried out with a set of agents that contact the server go periodically, so that when there is a task that must be carried out by an agent in particular, which meets the expected configuration, is assigned to it. When an agent executes a task a sandbox directory is created, on which all materials are updated associated with it, such as the content of a software repository.

Pipelines are the core of Go CD performance. A pipeline is made up of stages which are executed sequentially. Each of these stages is composed in turn of a set of works that can be carried out concurrently.

When all tasks at the same stage have been completed successfully, this task will be considered step has been successfully completed and then the next step defined in this step can be executed.

Pipeline

Pipelines can be triggered when there is a modification of an SCM, or when another pipeline has successfully completed, or may even depend on both at once. This server provides a notification system based primarily on email, means by which, the different members of the project of the states reached by the different tasks of the continuous supply process. For example, that it is over correctly a phase, a concrete test has failed, a manual phase has to be passed, etc.

In addition, this tool can be extended through a system of plugins, although it is possible to say that many have not yet been developed, so integration with external tools is right now poor.

One of the main advantages that can be highlighted from this tool is the way in which is presented visually all the information, since it allows to see in a clear and concise way how things really work. Provides a view representing the flow map of value (from the English Stream Map Value), which allows visualizing how a change in a repository concrete, causes the execution of certain pipelines. Therefore, if something fails it is possible to see the cause it triggered the error or effect caused by a pipeline in the following pipelines. This allows to have at all times a high traceability of what has happened.

Go CD also provides its users with a dependency management system with many options, making use of what is called material. A material is the element, pipeline or SCM, which depends on the triggering of another pipeline, so that it can build workflows complex. It is worth emphasizing a concept called Fan-in which refers to a dependence of N to 1, and that is contrary to the concept of Fan-out, dependence of 1 to N.

Jenkins

Jenkins [9] is a continuous integration server run by CloudBees and written in Java, which derived from the Hudson project. It is distributed under MIT license, and allows to carry out the following actions:

- Continuous construction and automated testing of software projects.
- Monitoring the execution of external services.

It is often used for projects written in Java, but can also work with other languages like C, Python, Ruby, etc., also allows to use a large number of construction tools, such as Ant, Maven, Kundo, Gradle, etc.

The Jenkins server obtains the information derived from the

construction faults and allows to send, through its system of notifications, this information to the members in charge of the project.

This system stands out for the number of channels by which notifications can be sent. In the reference [9] the following notification channels are shown: Android, e-mail, Google Calendar, IRC, XMPP, RSS and Twitter.

Unlike other continuous integration servers, it is not necessary to make a complex installation, but to use it simply a WAR file must be deployed. It also allows to use a lot of SCMs, some of them are CVS, Git, Bazaar, Integrity, Mercurial, Perforce, Subversion, Vault, etc.

To extend its functionality has a large number of plugins. These plugins allow you to integrate this server with other tools like IceScrum, more construction tools and a great number of test tools like Selenium, JMeter, xUnit, MSTest TRX, JSUnit, etc.

On the other hand, one of the extensions of Jenkins more striking for the present work, is the plugin delivery-pipeline-plugin, which provides a pipeline system, similar to Go CD, which allows to carry out a continuous delivery process.

It may be noted that Jenkins is one of the most complete continuous integration servers, which stands out due to its high functionality, integration with many environments and tools, an extensibility, which as described in the document is plugins and by the amount of documentation that exists of this tool. In turn this server is endorsed by a large community.

The decision to use the Go CD tool has finally been made. The alternative has been discarded of Jenkins because this server was created as a seamless integration server with an extension that allows to perform a continuous delivery function, while Go CD is a fully oriented server and created for continuous delivery, which can see the advantages outlined in the above table. Despite this, Jenkins has some advantages that Go does not have a large community that develops a large number of plugins, which provide an integration capability with a large number of tools. Using Jenkins would mean greater difficulty and carry out a continuous delivery task compared to Go CD understand a more tedious and elaborate configuration, using more plugins and scripts to solve in a cumbersome way certain situations.

DESIGN OF THE PROCESS

This section of the project will describe the design of the automated process that described in section 2.1., trying to make this process as similar as possible to the described by the Deployment Pipeline, and thus facilitating a continuous delivery. In first place a small study of the phases that will compose this process will be carried out, in secondly, the set of tools used in each phase will be decided, then a grouping of these phases in pipelines that can be implemented with Go CD and finally, researchers will detail how these pipelines have been made, each of the phases and the interconnection between all these elements.

Phases of the process

In general terms, the process that is to be implemented will have the phases collected in this diagram:

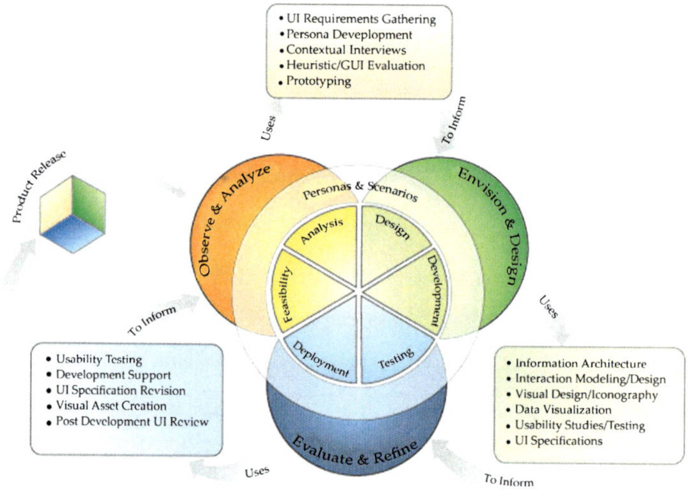

Figure 3: Phase diagram of the designed process

Figure 3 shows how the change by developers on a software repository will be responsible for triggering the execution of the automated chain, whose phases are, in general lines:

1. Unit tests: At this stage, the unit tests on the software, to decide if it meets the expected results. If they are correct, the next logical step is to build the project, so that will

automatically launch the construction phase. If the tests fail, it means that the last integration in the repository has an error, so the execution of the notifying process of the error.

2. Construction: This phase is responsible for building the software project generating a software artifact that can be tested and deployed in the right environment. If the project is built correctly, the next phase will be launched automatically, from otherwise the execution of the whole process will be aborted.
3. Deployment: In this phase it is about deploying the artifact generated by the previous phase in an environment where it can be tested, again if everything ends in a satisfactory way the next stage will be automatically launched.
4. Functional tests: Functional test cases are executed here, which have been previously designed by the test team, about the application that has been previously deployed in an environment prepared for the test phases.
5. Load Tests: This phase is responsible for running a battery of load tests, previously designed by the test team, on the artifact deployed in the setting prepared for this.
6. Replenishment: Once both the functional and load tests are completed, the software is ready for production, so the next step that is executed consists of replicating the software in the test environment and notifying the operations that have completed all the tests.
7. Production: This phase will be the only one of the process that will not be launched automatically, once the whole process is over and the operations team is advised of this, the team must access the server and, in view of the results obtained, approve the execution of the start-up phase with the press of a button.

Tool integration

In order to implement this process, the tools described in Chapter 2, a summary of the relationship of tools with the implemented process.

It can be seen that the Git tool was chosen as a version control system [10], where developers will make changes to the software project. On the other hand, for the unit testing part will use the already described framework JUnit, this will be used at this stage to test the changes made to the software project.

Maven will be used for the construction and packaging of the software, which has been tested in the unit testing phase, with the aim of preparing it for later deployment.

For the deployment and retraction phases, the Apache Brooklyn tool will be used. Responsible for deploying the newly packaged package, deploying the infrastructure necessary in the test environment to be able to carry out the subsequent test phases. These test phases are those of functional tests and those of load tests, in which will use for the development and execution of these tools the Selenium IDE and Apache JMeter respectively, allowing to obtain operating parameters of the application.

As for the phase of putting into production it should be noted that it is possible either to use Apache to do the deployment on the production environment or package the package with Maven and upload it to a repository.

Finally, the most important tool that will allow to integrate all the previous ones to carry out the entire process of continuous delivery is going to be the Go CD system. Once all the tools have been chosen and the phases that make up the process defined, the following step consists of dividing and grouping these phases into the building units that provides Go CD, which were explained in Chapter 2.

Grouping of the phases of the process

It has been decided to group the phases into three pipelines, each of which represents a concept logical process, in turn each of the phases correspond to a stage, so finally, the grouping is established.

There are three pipelines:

1. Build pipeline: Pipeline in charge of realizing the construction of the project and generating a package. Whenever you decide to package the software developed

in a package is necessary to carry out the execution of the associated unit tests, so it has been decided that these two parts are part of the same pipeline.

2. Tests pipeline: Pipeline in charge of performing the tests on the software deployed in a suitable testing environment. Therefore, the stages, coincident with the different phases, are the deployment of the application in the environment, the functional tests on the application deployed, load tests on the application deployed and once finished all these stages the fold the application.
3. Production pipeline: Pipeline that will only contain a stage that will be in charge to carry out the task of putting into production.

The following section will detail the design of each of these three pipelines and in turn of the internal phases that compose them.

Design of pipelines

First Pipeline

This pipeline will be in charge of the construction and packaging of the project so that in the subsequent pipeline is possible to deploy the application over a test environment with reason to run the functional and load tests.

This first pipeline should automatically check changes to a Git repository, so that when a change is detected on this repository, the first stage, which is responsible for the execution of unit tests, so the Git repository must be a material on which the pipeline build depends.

Whenever there are changes to the code repository, the version of the project will be updated software that is stored in the sandbox of the pipeline execution agent node, the project in the case that occupies this work is a Java project that will contain the tests together with the code.

Unit tests

The unit tests are the ones in charge of proving that the operation of the modules the project, in this case the methods of the various behave as expected, making sure that code parts do not contain errors and that you can start with the integration phase of these to carry out the automated execution of these the Maven tool has been used, making use of its command mvn clean test.

Once the unit tests are run, it is possible to obtain two results:

- The tests give a positive result, in this case the execution of the packaging.
- The unit tests fail, in this case the results of the tests will be notified the developer who made the upload to the repository, which allows the developer knows that your change has produced an error and act accordingly to solve.

In order to carry out the notifications, it is necessary to configure the Go CD security, so that there is a system of users, and in addition must configure the information of the server of mail, which will be in charge of sending the mails. This configuration process comes specified in Annex D.

Packaging

The packaging phase allows you to generate an application in the form of a package containing all the executables of the application, all the necessary resources and all the libraries of which depends on its execution.

Again the Maven tool has been used to carry out this packaging process, since it allows to satisfy the libraries indicated in the POM file of the project of way automated. The command used for this has been MVN package. If the application packaging successfully completes, the next execution will be triggered pipeline if the packaging execution fails, the process will end and you can see the reasons for failure on the Go CD server.

Second Pipeline

The next step in the chain consists of deploying the generated package in an environment of tests where they can execute on it the functional tests and relevant tests of load.

It can be seen that in this case the pipeline depends on several materials, for one side the construction pipeline, which should trigger the execution of the first stage of this pipeline, as long as the packaging stage finishes correctly, and on the other hand two Git repositories, the functional test repository and load test repository.

Deployment

This stage is responsible for deploying in a test environment the newly packaged. It is appropriate that this testing environment be as close as possible to the production to alleviate possible errors that may appear afterwards, when the application this one operating in the real environment, due to the difference of scenarios.

In order to carry out this deployment phase, and the subsequent withdrawal phase with its consequent monitoring, the Apache Brooklyn tool will be used. Because there was no way to make interacting, in a reasonably comfortable way, Go CD with the server Apache Brooklyn has been necessary to develop a plugin that extends the functionality of Go CD.

Development of a plugin

Plugins are the primary way to extend the base functionality that Go CD provides. The architecture of Go CD plugins corresponds to an OSGi architecture [11], thus trying that the Go CD application is the most modular, configurable and easily extensible by creation of extension points. Plugins are developed in Java language, language in the one that is implemented Go CD.

Extension points are defined as a set of Java interfaces and classes that abstract specific aspects of the application of Go. Plugins can be developed against these extension points by implementing interfaces or by extending abstract classes.

Go CD will convert the OSGi bundle to the plugin, which is packaged as a JAR file, and load it at runtime.

The architecture of plugins is roughly composed of:

- Infrastructure Go Plugin: Module responsible for responding to changes in plugins, to convert plugins into OSGi bundle and load them.

Activator Go Plugin: Module responsible for loading all classes in the plugin that are not in the dependencies directory and register services for the extension of plugins. It is an implementation of OSGi

BundleActivator.

- Api Go Plugin: A module that contains the APIs of all extension points.
- Api Go Internal Plugin: Module containing the APIs for a set of services that will be available for the plugin at runtime. For example the logs service, LoggingService.

For more information on the architecture used by Go CD see the reference [8].

In the current version of Go CD, there are several extension points defined by the Go CD team:

- Package repository extension as material.
- Extension of tasks.
- Extension of notifications
- Extension of SCM

Functional testing

The functional tests are those tests that allow to test the software developed in execution to validate exhaustively if the software works as it should and if it complies with those requirements that have been specified during the planning stage.

The tool that will be used in this part, both to create and to run the tests, will be the Selenium framework. The test team will be in charge of designing and generating, with this tool, the set of functional tests to run and import them into a Java project.

For this case study, a Git repository will be assumed, in which the test equipment will be responsible for uploading Java projects with functional tests. Therefore, this pipeline must have a dependency with the Git repository containing the tests. The Go server will be in charge of checking if changes occur on this repository and update the sandbox of the agent with the project functional testing, however, unlike other dependencies, this will not trigger the pipeline, but will only serve to have all the functional test reviews.

For the later execution of the Java project, which contains the functional tests, the Maven tool, responsible for executing them as if it were a unit test, using of Selenium webdriver, in charge of interacting with the browser and running the tests. A once the tests are executed, some reports will be generated with the results, and if the tests fail the execution of the automated process is finished. If the tests are all correct, the execution of the next phase, execution of load tests.

To carry out the tests it is necessary to install a service that provides an X11 server dedicated in the agent node that will be in charge of its execution.

Load tests

Load tests are a type of performance testing that attempts to measure behavior of an application when it receives a certain number of operations, for example a number of users who use the application concurrently and make x requests on it at the same time.

Getting different parameters of the application, such as the neck of bottle of the application. Again, as in the case of functional tests, the test team will be responsible for design and generate a set of load tests that will be uploaded in a Git repository. Is by that another dependency of this pipeline will be this Git repository. The Go server will be in charge to review if changes occur on this repository and update the agent sandbox with the load test project, as in the previous case, this dependence will not cause the pipeline but will only serve to keep the load tests. In order to carry out the creation and execution of the tests, the JMeter tool.

The link between Go CD and the JMeter test tool will again be Maven, but in this case it is necessary to use a plugin that allows to execute the tests and in turn generate some reports with the results in an HTML format. Once the tests are run reports are generated which will be available from the Go server. At the end of this phase, the set will be finished necessary to collect data to know if the application is deployed in a production environment.

Fold

Once the test phase on the application in the test environment is finished, the next step have to replicate it from this environment. To do this again will use the developed plugin during the work, but in this case, in its functionality of withdrawal. Once this withdrawal is carried out, it is necessary to inform the operations team that carried out the test pipeline correctly and that there is a new version of the application ready to replicate. For this purpose it has been decided to make use of the system of notifications of Go CD, notifying of this event and sending information of which project review fired the process and proofreading, both load and functional, have been executed.

The operations team is now in charge of reviewing the results of the tests on the Go server to decide whether to carry out the production release of this version or do not.

Third Pipeline

Finally, the third and final pipeline has as its main mission the production of the application that has been processed by the previous pipelines.

Satisfactory, the operations team receives a notification with the necessary information to know that there is a new candidate version to be put into production. Unlike rest of pipelines, it needs the manual approval of a team member to start its execution.

Production

The last link in the chain of the software delivery process consists of putting the software in production. This phase is one of the most critical phases of the software process, because a failure of the application in the production environment can result in a large cost for the parties involved.

Facilitating a manual execution button provides the advantage that you can analyze all results obtained from test phases before choosing whether or not to carry out the phase of launching of the application, ruling out possible false candidates for the production. This is the characteristic that differentiates the proposed process to carry out continuous delivery, in which it is tried to frequently put the software in a prepared state for the putting into production, of the model of continuous deployment, in which this phase would shoot without any manual approval.

In this stage of production, two options have been considered: first, to deploy the application in a production environment making use of the Apache Brooklyn tool, and the second, which is the option chosen in this implementation, pack the application and load it in a launch repository, making use of the Maven tool, for its later deployment.

CONCLUSIONS

In this chapter researchers intend to present the conclusions derived from the present work. For this purpose, researchers will review the initial objectives of the work and the degree of achievement of these, a valuation of the contribution that has supposed on a personal level and as a last point will be proposed a set of possible extensions and improvements.

As described in the introduction to the section, the fundamental objective is to demonstrate that you can carry out a complete automation of the life cycle of an application, integration of existing tools, so that implementation of a continuous delivery model.

Following the study that has been detailed in this document, both in the design and in the evidence, it has been shown that the proposed objective has been met. It has been possible to put in context the set of tools chosen by carrying out an assembly of each one of the tasks allowing you to process an example application from its development to its preparation for production, also facilitating a rapid feedback of the results of evidence and the status of the application to all members of the teams involved in the software creation process through the notification system provided by the server

CHAPTER 2

AUTOMATION OF WEB COMPATIBILITY TESTS IN A CONTINUOUS SOFTWARE DEVELOPMENT ENVIRONMENT

Ivan Stanimirović

The following section provides a brief theoretical framework on web compatibility tests. Then, the proposed technique and the algorithms used and the results of the implementation of the technique are shown. Finally, book outlines the conclusions and future work.

WEB COMPATIBILITY TESTS

That a website is compatible with all browsers means that it looks the same (or very similar) in all of them. Some authors consider it sufficient if the site can be perceived by the user with the same characteristics, in browsers such as Internet Explorer, Firefox, Chrome, Opera, Safari and

Mozilla. The problem is that not all browsers interpret HTML code and style sheets (CSS) in the same way [11]. Some of these differences are so important that they cause the malfunction of the site or the loss of display. One of the goals of building a website is that it can be visited by the largest number of people (and that they see it correctly). Therefore, it is very important that the site works in the same possible browsers [10]. Compatibility Tests or Cross-Testing [12], are tests that are performed on a particular application by checking its compatibility with all Internet browsers and market. For this, there are different techniques:

· Verification of compliance with standards (such as W3C or ECMA): It consists of analyzing the graphic components of the website in different browsers, to verify that they follow the guidelines and specifications of the standards. This technique is preferably used in intermediate stages of the development process, it being understood that the actual design will be carried out with the consideration of corresponding standards [13], [14].

- User interface testing (UI): It is the most common of the techniques. Can be performed manually or with specialized software (automated testing). The purpose of this type of test is to check the visual content of the website at through the navigation of its pages in the different browsers [15], [16].
- Analysis of the Document Object Model (DOM): It is a dynamic technique which is to compare the behavior of a web application in different navigators, identifying the differences as defects. The comparison was by combining structural analysis of information in the DOM together with a visual analysis of the page in question. [17].
- Image Comparison: This technique is based on taking a screenshot of the site in one type of browser, and compare it with another site capture in another browser other than the first. If both images match, then the site will be compatible between both browsers [18].

Of all of them, the technique of comparison of images is chosen by diverse authors [18], [19], [20], [21], [22] as the most suitable for testing web compatibility, due to the ease of launching it

PROPOSED TECHNIQUE

In a continuous development environment, automated functional testing play a key role, since they allow rapid regression in small development cycles for each change introduced in the product of software. With the emergence of continuous integration servers, it is possible to detect the introduction of defects in the source code as soon as possible [23]. Of all incompatibility problems are detected by performing manual exploratory tests on different browsers. These teams often rely on specialized software to perform comparison of pictures.

At present, there is a large number of non-proprietary tools that allow to implement algorithms for image comparison [24]. However, the problem of these tools is that none can be coupled to the execution of tests functional on websites [19].

The proposed technique is, therefore, to complement the testing process functional on a website, with an algorithm of comparison of images automated, and in this way accelerate the detection of incompatibilities between different browsers.

The environment was developed using Java and consists of the following components:

- A tool for interacting with different browsers.
- A tool to verify compliance with conditions (existence of elements, expected behaviors, etc.)
- A tool for processing images.
- An algorithm for performing image comparison.
- A mechanism for running parallel tests on different Operating systems and browsers.
- A reporting tool to display the results of the performance of tests

The following facts were considered:

1. The tests are run through Maven. This tool is responsible to compile the code and give the command to the main node to begin its execution.
2. The main node is a hub of Selenium Grid, a server that allows to trigger test execution threads on multiple nodes in parallel. East Server instance browsers configured

on each node. The tests are generated using Selenium Webdriver.

3. During the execution of the tests, screenshots are made and stored on a hard disk to which all nodes have access synchronized. The tests conclude when all functional validations are completed. These validations are performed using TestNG.
4. The image comparison algorithm takes screenshots generated, and begins to analyze them in pairs generating other images results, which then stores them on the same disk where the catches are found.
5. When the process is finished, the final report is generated with the results of the execution of the tests in an HTML file.

The screen capture process

The interaction with the website is the responsibility of the Selenium tool Webdriver. The same test script is run in different browsers in parallel, through a configuration of Selenium Grid that allows instances of a thread of execution for each virtual machine (node) of the environment. As the tests go from one page of the site to another, it is taking a capture of screen.

One of the challenges was to find the most get different screenshots with the same dimensions. An alternative was to use a function given by the Selenium Webdriver tool to obtain screenshots of the site. But this caused problems with the catches of the Mozilla Firefox, where they not only contained what would be seen in the screen, but all the content of the page that is accessed by performing the scrolling-down / up. The second option was to use an API of Java, which allows taking screenshots of the computer. However, while the results were images with the same dimensions, they all contained the toolbar of each browser, and at the time of images compatibility tests always failed.

Finally, the algorithm is consisting of three steps:

1. Put the browser in "full screen" mode.
2. Get the full screen dimension using the Java Toolkit API.
3. Generate the screenshot through the Java Robot API.

Image comparison algorithm used

The images taken are stored on the hard disk of a computer, with a nomenclature that represents: the identifier of the test. The name of the page in the one that took the catch; the Navigator; and the name and version of the operating system of the node.

The tool to carry out the verifications of the expected results and (TestNG), presents a mechanism for executing instructions upon completion tests. This mechanism is defined by AfterSuite / AfterTest / AfterClass. It also allows the execution of pre-test instructions

To perform the image comparison, the @AfterSuite annotation is used. Within the method that contains this annotation, the call to the algorithm of comparison of images.

The first step in this stage is to get all the images corresponding to a certain node, using the browser and operating system specified in the name of the images. Then, the images are obtained corresponding to a second node. Once the images are obtained corresponding to the first two nodes, an algorithm that takes an image of a node and searches for its corresponding (according to the identifier and name of page) between the captures of the other node. Comparing images begins to be performed in pairs.

Finally, the image comparison algorithm is executed:

1. It is verified that both images have the same dimensions. If they are not equal, the comparison is ignored and a fault is obtained
2. An array of pixels is obtained from the base image, and researchers begin to traverse all pixels one by one, comparing them with their reciprocal in the other image. If they are different, the position of the different pixels is saved.
3. If there were no differences, the compatibility test has been completed with success. In the event of a fault, a third image resulting from the comparison is generated. The image is painted with dots in each position corresponding to the different pixels. This image is saved with a heat map format next to the pair of catches that were analyzed.

Finally, the result of the comparisons can be observed in the final report. The final report is generated with ReportNG, as a complement to TestNG. For the compatibility testing section, an HTML report has been developed that has the format shown in Fig. 7 which is accessed by clicking on each fault of the comparison.

RESULTS OBTAINED

This technique has been implemented in a real environment of continuous development, and the results demonstrate their efficiency. On the one hand, the implementation of the algorithm was a very simple task to be carried out by the development team. The technique has accelerated by 92% the time of execution of the tests of compatibility and therefore the whole test stage (see Fig. 8). A conventional compatibility test performed by a team member takes approximately 10 minutes depending on the page. A compatibility test on the same page with the technique proposal takes approximately 1 minute, required to observe the heat map with the results. A decrease in total time of the release process of each version of the site, from the implementation of the proposed technique is observed.

CONCLUSION

The proposed technique is an initiative to automate the compatibility through an image comparison algorithm. To validate it, it has been implemented in a large company that works with continuous development of software. The results show that the tool accelerates the testing, through the automation of web compatibility tests.

On the one hand, test execution times decreased by 82%, and this also reduced the total time of the release process of the site versions. In addition it allows to generate reports in HTML format very easy to understand.

With the implementation of this approach, comprehensive visualization of components in search of incompatibilities about the different browsers. In this way, manual tasks are simply reduced to observe the reports. Likewise, it allows to detect more defects that can be passed when the tests are performed manually. However, there were drawbacks in comparing sites that contain different advertisements and / or emerging elements, producing false in the tests.

As future work it is proposed, firstly, to improve the algorithm implemented in order to avoid comparisons in non-site elements web (such as ads). Secondly, tools will be used virtualization (such as the Docker tool) to running functional tests. Finally, techniques that can be implemented to avoid work required to observe the results of way to make the process fully automated. Also, algorithm to avoid alarming scenarios should be investigated because of small differences between the browsers.

CHAPTER 3

CONTINUOUS SOFTWARE DELIVERY IN A CLOUD SERVICE

Ivan Stanimirović

The development process is a key factor in the viability of a project since it affects both its planning and execution. However, it has a special relevance when it›s a business project. To get a product to market, in special software-based products, it is necessary to coordinate very well all the components involved in its development. For this reason you should follow a development methodology containing tools, models, and methods improve the development process without forgetting the quality of the software. This methodology it will allow to structure, plan, and control the process of development by enhancing the process.

As the product is developed there are tasks that are increasing in complexity, to ensure the viability of these tasks should automate them. In this way you will controlling the increase in complexity. It is also important to automate the most recurring tasks, such as tasks of testing and start-up of production of the product. Of this way the development team focuses its efforts where it provides the maximum of value: developing new features, write new test or by improving the put into production. For this reason a tailored development pipeline will be developed the continuous delivery model. The pipeline aims to automate all the recurring tasks des from development to deployment in different environments, it will also allow frequent delivery of new versions of the software.

Cloud gaming platform provides a service where a user can interact with applications running on the cloud, in this case games, from any device connected to the internet.

INTRODUCTION

A development methodology is a guide used by a development team to Please and join efforts to achieve a common goal. Currently in the community, there is a wide variety of methodologies (such as development in cascade, development spiral, iterative design and incremental development). Each of them with properties well differentiated in function of the casuistry you adapt.

It should be noted that both the absence and the implantation of a methodology not appropriate, are counterproductive and will directly affect the performance of the team of development. For example, in the absence of a methodology, integrate the work which will developing each component of the team it would be very complex, lengthening the time that it takes to deliver a new version of the software. This problem is going to go in increase as it grows the team either the complexity of the product that is developed.

Continuous delivery

The continuous delivery consists in developing a process of implementation in production allow be able to release new versions of the software frequently. At present in the community, the application of the continuous delivery does not follow any established standard and is carried out by means of manual integration of the different systems involved. These systems depend on the concrete implementation.

The application of the continuous delivery has the following advantages:

- Early detection of problems and potential points of failure.
- Rapidity and simplicity in production of new versions.
- Need fewer human resources dedicated to tasks that are automated.
- Development of know-how about the process because the iteration on it.

To develop pipeline comprises both new versions of the software and its dependencies, such as the provisioning of servers, updating and management the configuration of them.

The motivation

When working in a startup environment is gives me the opportunity to learn a lot. Since the restrictions are minimal in this environment as to which technologies use, have great flexibility to use the latest technologies.

In addition, the task of applying a methodology in a multidisciplinary team is something new since you must coordinate in the field of operations and systems, with the development of the product.

Integration between systems is a pending matter, and there is that figuring how to get systems to understand each other and share results. There is also a trend to the rise of cloud systems based on Microsoft Windows to which the current Open Source tools just begin to support partial, while these tools are being used without any problems for clouds based on Linux. For these reasons, this project represents a technological challenge.

Bring value to the company where work is a key aspect in the motivation for the realization of this project, as is going to apply in an environment productive. As student is very motivating to work in a solution that has to adapt to the features of the real world.

It is very motivating to work in an environment in a company that develops a product in order to market it worldwide. On the other hand, it is interesting to get systems to work in the service of people to facilitate the work. By automating recurring tasks people can devote to more interesting tasks where they can really develop their capabilities.

Analysis, search and implementation of a development methodology

Currently many firms do not apply a clear development methodology. For this reason the first part of the chapter consists in the study and analysis of the State of the art of the development methodologies with the aim of incorporating the best of them using a framework adapted to the needs of the development team.

Some frameworks are characterized by the agility with which allow the development of the software product. In the case that concerns us, this feature is critical since the platform is in continuous development. For this reason, the second part of the project is to implement the pipeline of software development that fit the model of continuous delivery, and therefore allowing to maximize the agility with which works the team.

As a result of this chapter, will be an improvement in the development process of software in a real case and a process of continuous delivery that will be implemented researchers can automatically deploy new versions of the platform in the cloud.

A goal is to find a methodology of software development that suits the characteristics company and product that is developed. Also provide the components of the company that adapt to the new methodology and go by retrospective meeting to go adapting the methodology each time better to the needs of the team of development.

Also one should look and set up the necessary tools that allow to the equipment components follow the methodology of development.

Among requirements, the framework associated with the methodology, as well as processes and tools have be scalable. In this way may be adapted to a potential growth in staff of the company, or to the increase in complexity of the product that is developed.

Implementation of the software development pipeline

Automate the process of deployment of new versions of the platform in the different environments (development and production), this is implemented a pipeline of software development following the model of continuous delivery. Also develop to return to the stable State in case of failure during the deployment mechanisms. By last to also develop a mechanism for reporting the status of the different environments at a given time.

Prepare the automation of the deployment of the software so that it is independent the characteristics of the environment. For example employee hardware or configurations necessary for a particular environment.

On the other hand updates in production must be the least disruptive possible about the service in March, in the time of non-availability of the service.

There are two major tasks that correspond to the objectives of the project. Each of these can be subdivided in the tasks shown.

METHODOLOGY OF SOFTWARE DEVELOPMENT

In this chapter the methodology of software development will be introduced to continue the development team. Initially shown a brief description of the concept of software development methodology, then explain the State of the art, an analysis of frames will be of software development current and past, will be presented the methodology chosen along with its initial configuration for that is fit the most to the particular needs of the case.

Definition of software development methodology

Software development methodologies are a set of procedures and techniques that help to structure, plan, and control the process of product software development.

By introducing a simile, the procedures are as a book of recipes in the kitchen, in that is an indication step by step activities to achieve software desired, indicating also the people who should be involved in the development of the activities and the role they will play. Also details the information necessary to start an activity and information that should occur as a result of the same.

Techniques indicate how a given activity should be performed in the methodology. They combine the use of models or graphic representations together with the use associated detailed procedures. It must be considered that a particular technique can be used in one or more activities of the methodology of development of software.

Each methodology is based on a philosophy of development which determines how will be the phases during the development and the relationship between them. For example, the methodology Determines if the testing of software is included in the phase of development or on the contrary the test has an own stage in the process.

State of the art

There are different methodologies for development according to the type of project for which they are suitable, stages comprising the methodology, the relationship between such stages or how to make the collection of requirements [15].

The most common development methodologies are cascading, the prototyping, development iterative development and incremental design, the development spiral, and agile development.

Waterfall development is the methodology that rigorously ordered stages the process for the development of software, in such a way that the beginning of each stage should wait for the completion of the previous stage.

The main stages of the waterfall model is the analysis of requirements, design of the architecture and the program, implementation, verification and maintenance.

Requirements analysis

This phase analyzes the needs of the end users of the software for determine what objectives should be covered. This phase emerges a memory called document specification of requirements (SRD), which contains the specification full what researchers should have the software without entering internal details.

It is important to point out that at this stage is must agree on all what is required of the software and is that what will follow in the next stages, not being able to change to half of the software development process.

Architecture and the program design

At this stage it decomposes and organizes the software in elements that can be separately, taking advantage of the development team. As a result, it appears the design document (SDD) software, which contains the description of the structure relational software and specification of what to make each of its parts.

It is convenient to distinguish between high level design and architectural and detailed design. The first is intended to define the structure of the solution (after a phase of) analysis described the problem) by identifying large modules (sets of functions going to be associated) and their relationships. This is defined the architecture of the solution chosen. The second defines the used algorithms and organization of the code for beginning the deployment, compliance with the requirements of the user as well as also the analysis needed to know what tools to use at the stage of implementation.

Implementation

This is the phase where implements the code source, making use of prototypes as well as test to correct errors. Depending on the programming language and its version libraries and reusable components within the same project are created to make programming a faster process.

Verification

The programmed elements are assembled to compose software and verified that it works correctly and that complies with the requirements, before being sent to the final user. Once finished the verification is when is the delivery of the software to the user final, which starts to be used for the first time.

If not verification, becomes the stage of requirements as if they were an extension or improvement of existing software.

Maintenance

The software maintenance is the stage where the user detects the extent to the software is meeting their needs and leads to the implementation of new features again initiating the process of development from the first stage. Waterfall development has the advantage that the documentation produced to measure that the different phases are completed one after another. Even so, actually a project rarely follows a linear sequence and, for this reason, it is usually to create a bad implementation of the model.

On the other hand, the process of creation of the software subject to this model is expensive in time, since up to that software is not fully complete is not provided to the customer. Like this also, any error detected in the verification stage design leads necessarily to the redesign and new implementation of the affected code, increasing from this way the development costs.

This model is inflexible to changing customer requirements, as well. Therefore, only conforms to projects where the requirements are very clear and are highly unlikely to suffer changes.

In the case of the product type, requirements are small variations frequently (for reasons of business, technological restrictions is it detects to the first tests, etc.). On the other hand, the product that is developing must always be available for demonstrations before the potential customers. For these reasons, the model has been dismissed to be implemented.

Prototyping

The prototyping-based development is a model of evolutionary development. The prototype should be built in a short time, using appropriate programs and without using many resources.

Requirements Planning

Users, administrators, and other it staff discuss and agree the needs of the business, the scope of the project, restrictions and requirements of the system. This phase It ends when the team agrees on the key issues and obtains authorization management, who makes the decisions, to continue.

User design

During this phase, users interact with systems analysts, taking part in the development of models and prototypes that represent all the processes of the software, inputs and outputs. Designers often use a combination of development set of technical applications and tools to translate the needs of the user in working models. The user design is an ongoing interactive process that allows to users understand and finally approve a software design that meets their needs.

Construction

It focuses on the task of programming the software. However, users are still participating and still may suggest changes or improvements to develop the screens or actual reports.

Cutover

In this phase is data conversion, testing, change to the new software and the training of users. In comparison with traditional methods throughout the compressing process. As a result, the new software is built, delivered and operation long before.

Prototyping-based methodology focuses on the representation of aspects the software that will be visible to the customer or end user. This design leads to the construction of a prototype, which is evaluated by the customer to a feedback; Thanks to this, the software requirements are refined is It will develop. The interaction takes place when the prototype is set to meet the customer's needs. This allows that at the same time the developer better understand what to do and the customer to see results in the short term.

In this way, the prototype is built and serves as a mechanism for the definition of requirements. So it is usual that the prototype, in part or in its entirety, is discarded.

Should take into account that when the real development is carried out shall be given in the quality. For example, if developed in a certain programming language by quickly, at the time of developing real programming language will be scanned is suited to the project.

This model conforms to cases where, for the reason that is, is difficult to get an idea what the customer needs. So is built a prototype for the client see a possible solution and begin to work on this.

It is not difficult to imagine the main requirements, so no sense wasting resources making requirements of this caliber.

Iterative development and incremental design

The main idea behind this methodology is to develop incrementally, allowing the developer to take advantage of what researchers have learned along of the previous development, increasing deliverable versions of the software. Learning it comes from two aspects: the development of software and the feedback provided by the client after the use of the software. The key steps in the process are to begin with a simple implementation of the requirements of the system, and iteratively improve the sequence of evolutionary updates until the full software is implemented. In each iteration makes changes in the design and added new features and capabilities to the software.

The process consists of two stages, the initialization stage and the stage of iteration. As shown, the initialization stage creates a first version of the software as soon as possible and at the stage of iteration is carried out planning, requirements, analysis and design, implementation, deployment, testing and evaluation.

Initialization stage

Creates a first version of the software. The goal of this stage is to create a product with which the user can interact and, therefore, feedback process. It must offer a sample of the aspects key of the problem and provide a solution as simple enough to be understood and implemented easily. To guide the process of iteration is created a checklist of project which contains a history of all the tasks that need to be carried out. It includes aspects such as new features to be implemented and areas of redesign of the existing solution.

This checklist is reviewed regularly and constantly as a result of the phase of analysis.

Iteration stages

This stage involves the redesign and implementation of a task from the list of control of project and the analysis of the most recent version of the software. The goal of the design and implementation of any iteration is to be simple, direct and modular, for power support the redesign of the stage or as a task added to the list of project control. As in many cases the code represents the major source of documentation of the software, analysis of each iteration is based on the feedback from the user and the analysis of the available features of the program. Involves the analysis of the structure, modularity, usability, reliability, efficiency and effectiveness (reach the) goals). The project control list is modified according to the results of the analysis.

The implementation and analysis are guided by the following rules:

Any difficulty in design, coding, and testing a modification is it will solve redesigning or recoding.

Modifications should be easier to do as they progress the iterations. If it is not, there is a serious problem caused by weak design or by excessive proliferation of patches to the software.

Patches should normally stay only for one or two iterations. They are necessary to prevent the redesigning during an implementation phase.

The existing implementation should be analyzed frequently to determine what such conforms to the goals of the project.

The client's opinion should be requested and analyzed to indicate deficiencies in the implementation.

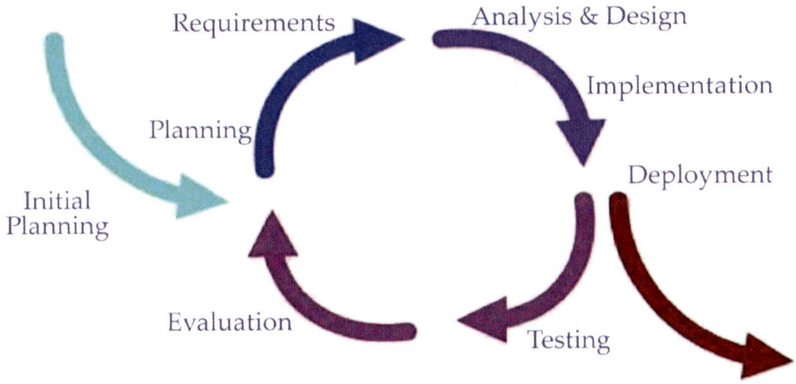

Figure 3.1: lifecycle of iterative development model and incremental design.

An advantage of this model is that the risks are diminished since customer reviews regularly carried out work and thus, can verify that it really is what you need.

Another advantage is the possibility of an easy way to change the requirements as they can be modified between iterations. Also, costs therefore if any resulting functionality of an iteration is not the taste of the customer and goes to the version above, just lost the resources allocated to an iteration. Finally, to the end of each iteration the client has a version of the product functioning, so researchers don't have to wait to the end of the development process to start using the product.

This model fits quite well to the product that is developed. On the other hand part, it is a way of working that suits the style of work of the team of development. For the moment, it would be the most suitable methodology for the analyzed.

Spiral development

The spiral development methodology takes into account the risk that appears to the develop software. So begins looking at the possible alternatives of development, more manageable it opts for the risk and the spiral cycle is done. If the customer wants to continue making improvements in software, turns to evaluate the different new alternatives and risks and is another turn of the spiral, so until it reaches a time in which the developed software product is accepted and do not need to follow improving with a new cycle.

It basically consists of a series of cycles that recur in a spiral, starting from the Center. Is often interpreted that within each cycle of the spiral is following a model cascading. Spiral development can be seen as a model evolutionary it combines the iteration and the waterfall model, with risk management.

Cycles

In each turn or iteration should take into account the objectives and the alternatives. The goals mark the needs which should cover the product. The alternatives, on the other hand, They mark the different ways of achieving the objectives of a successful and from different points of view, such as features or the risk assumed in each alternative.

For each cycle will have four stages: determining the objectives, analyze risk, develop and validate, and plan the next iteration.

Determine the objectives

At this stage the outputs are set to get as requirements, specifications of the software or user's manual. Also the restrictions are set and identify the risks of the project and strategies to avoid them.

Analyze the risk

The study of the causes of possible threats is carried out and probable unintended consequences of these. Alternatives are evaluated. Is it because have a prototype before you begin to develop and validate.

Develop and validate

Depending on the outcome of the risk assessment, a model is chosen for the development, which can be any of the other existing, such as waterfall, or prototyping. Thus for example if risks in the user interface are dominant, a model of appropriate development could be the construction of prototypes.

Planning the next iteration

If the result of the previous step is not adequate or is needed to implement improvements or functionalities are planned steps and begins a new cycle of the spiral. The spiral has a shell shape and is said to keep two dimensions, the radial and the angular. The angular dimension indicates the progress of the project from the software inside of a cycle. The radial dimension, on the other hand, indicates the increase in the cost of the project, already with each new iteration passed more time developing. This model is widely used in large, complex projects such as, for example, the creation of an operating system.

To be a model oriented to the management of risk, one of the fundamental aspects of its success lies in that equipment which apply it has the experience and skill required to detect and correctly categorize risks.

Taking into account the type of product which is developed, makes no sense go making risk analysis regularly, since although the software is complex and has a large number of features, the expenditure of resources on risk analysis not it would offset.

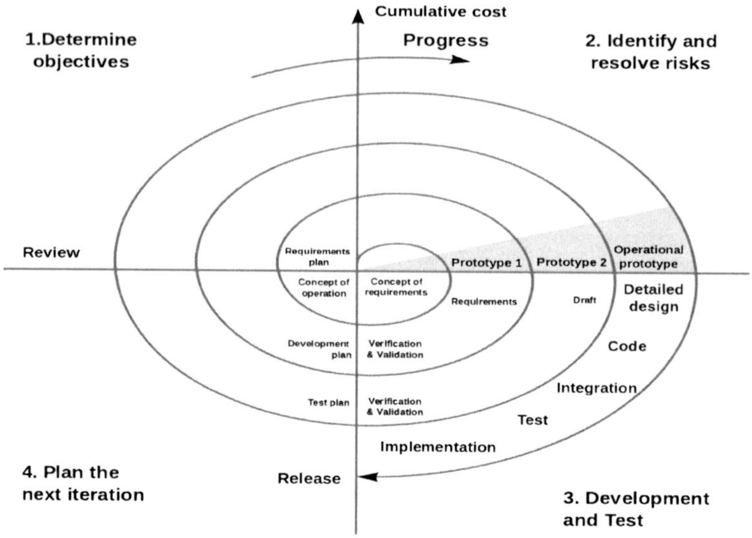

Figure 3.2: the development spiral model lifecycle.

On the other hand, to include the methodology cascade back to have the constraint that the requirements are not subject to change

Agile software development

Agile development is based on iterative and incremental, development where the requirements and solutions evolve through collaboration with organized auto groups and multidisciplinary [11].

There are many frameworks based on agile development, most minimizes risks developing software in short time periods. The software is developed in units of time calls iterations, which should last from one to four weeks. Each iteration includes planning where an analysis is made of requirements, the phase of collaboration where developers design, coding, review of the document the new features, and finally there are the delivery phase ending with the feedback provided by the client concerning the performed delivery. An iteration should not add too much functionality to justify the launch of the product on the market, but the goal is to have an increase in software running at the end of each iteration.

At that time the team returns to assess the priorities of the project. Agile methods emphasize face-to-face against making communications of extensive documentations. Teams must include testers, editors of documentation, designers of iteration and project directors. Agile methods also emphasize that functional software is the first measure of progress.

This model inherits the advantages of iterative development and incremental design model. Likewise, the way in which it is defined gives greater relevance to people that continues, which increases the motivation and involvement of the people and therefore the productivity of the team.

Agile development is suited to projects where the requirements are modified with certain regularly (even be added new often), where the features are developed quickly (without neglecting the quality) and subsequently improved.

The cost associated with each iteration is much less to follow a traditional model, in addition researchers are confident that the product adapts to the needs of the client, since this is a figure in the development process. Therefore, the model of agile development is suited to environments where the product is very alive, though are concerned of a software complex. Achieves high productivity at a lower cost. For this variety of reasons it will be the model chosen to implement this project.

ANALYSIS OF AGILE DEVELOPMENT FRAMEWORKS

Development frameworks are guides that explain how to implement a methodology concrete. Knowing that researchers are going to work with an agile methodology, researchers will review briefly the agile development frameworks.

Extreme Programming (XP)

This framework of development was very popular about the year 2000. It focuses on describing engineering practices such as Pair Programming and Test Driven Development. The big problem that has is to focus on the part of development makes it difficult to follow up for the management and also makes direct contact with the customer.

Lean Kanban

This framework is based on Lean Manufacturing, which are practices of production which aims to use the least possible resources and minimize the waste of the same, while maximizing the value of the product to the user. Read help to produce little complex software very quickly, focusing on avoiding meetings, unnecessary tasks or documentation.

The type of product which is developed would be difficult to manage according to Lean Kanban, is a complex software.

SCRUM

Scrum [16] is a framework for the development of a process-based software iterative and incremental used in agile software development-based environments. It defines a strategy for software development where the development team works as a unit to achieve a common goal. Scrum enables the team organize themselves through, for example, the promotion of everyday communication between its members.

A principle of Scrum is to note that during the development of a project the customers can change their idea about what they want or need (requirements) and the unpredictable changes are hardly acceptable if you plan everything from the first, as you would traditionally.

Scrum adopts an empirical approach, focusing on maximizing the capacity of the team deliver software that responds to the emerging requirements quickly. This intends to perform tasks in iteration, called sprints that can last from one week to one month. A sprint aims to deliver an increase in software that you are ready to move to production.

It has been decided to work with Scrum by the proximity of the development team with the requirements of the customer and its possible changes. It concerned the development of a product lively is a key factor for the election.

Since that it has been decided to work with an agile methodology and development framework Scrum is chosen, researchers will define it in depth. The main components of Scrum are roles, artifacts, and regular meetings.

This methodology defines a few leading roles and a few side. The roles main involved directly and secondary roles, even if they do not, should be taken into account that make possible the process of software development.

It is called to the set of principal roles team Scrum. These roles are here mentioned.

Developer Team (DT)

It is formed by all the members of the development team. Scrum defines that the team must be composed of 3 components like minimum and maximum 9. In a ideal setting all its members should be prepared to perform any task. In reality, these teams are usually interdisciplinary but with time the exchange knowledge and newly developing competencies in all components the areas covering the team.

The team is responsible for the Definition of Done (DoD), which is the definition of the State in which a task must be to be considered as made.

Product Owner (PO)

It is responsible for product figure from the point of view of the business, and represents the link between the customer and the development team. Perform requirements is one of responsibilities, as well as the prioritization of tasks to be carried out by the development team.

Scrum Master (SM)

It is the figure that facilitates the correct monitoring of the methodology. With the aim of lead the implementation of this methodology, as part of my project RESEARCHER played this role in the company. It should be noted that it is not a role as leader of the group, since the team is self-organized. This role does not have why require a dedication to full time, in a way that can be performed by one of the components of the team formed for this purpose. This scenario is very common.

Include secondary roles to customers where the socket is made of requirements. And managers, which although they do not form part directly from the methodology they make it possible. For example, hiring a new developer as they are necessary.

Artifacts

Artifacts in Scrum facilitate follow-up to the methodology. Researchers have the product backlog, sprint backlog and the burndown chart.

The product backlog is a prioritized list of everything that is necessary in the product and the only source of requirements for any changes that you would like to make to the product. The product owner is responsible for the backlog, including its content, availability and prioritization.

In this list researchers find all the features, functions, requirements, improvements and corrections which are the changes to be done to the product in future deliveries. The the backlog items have attributes like a description, an order, an estimate and a value.

As the software is being used by the client, the latter provides feedback and thus it makes the backlog grows both in size and detail.

Sprint backlog

The sprint backlog is a list of tasks to perform during a sprint, this list is a subset of the product backlog and indicates the hours be devoted him daily.

The following rules apply to the making of the sprint backlog at the beginning of the sprint: The members of the development team choose the tasks individually. The work is never assigned, there is a figure that assign tasks. If the work is not clear, define a task of the sprint backlog with a larger number of time and then subdivide it.

Burndown chart

It is a graphic that helps us to see the progress of the team during a sprint. By displaying the sum of the estimated time remaining (in days) on the vertical axis and the passage of the real time on the horizontal axis.

Using the burndown chart can be identified if the team will be able to cover the tasks of the sprint backlog during the sprint. For example, if half the straight Sprint of to perform actual work team is well above the ideal line, then one must negotiate with the product owner the tasks that must be eliminated from the sprint backlog to deliver an increase in software at the end of the sprint.

Meetings

One of the keys of Scrum is the coordination and communication between the people involved. To achieve this, specific meetings are defined. They are limited in time in all cases.

During a sprint, a meeting on the status of a project is done daily. To this meeting goes the development team and the Scrum master. This meeting takes place in the same site and at the same time, this is done so to avoid confusion and for provided that it is carried out. It should last 15 minutes maximum.

During the meeting, each team member answers three questions:

- What has done since yesterday?
- What happens up to tomorrow's meeting?
- Have a problem that has prevented him from achieving his goal?

It is part of the role of Scrum master remember these three questions. The daily meeting improved communication, eliminates the need to maintain other meetings, identifies and removes impediments relating to the development, highlights and promotes the quick decision-making, and improve the level of knowledge of the development team about the project. It also constitutes a key meeting for inspection and adaptation.

Sprint planning

The work to be performed during the Sprint is planned in the sprint planning. This plan is created through the collaborative work of the Scrum team to the full. The sprint planning meeting is restricted to a maximum of eight hours for a sprint for a month. For shorter sprints, the meeting is proportionately shorter. For example, Sprint's two weeks a meeting of four hours.

The sprint planning consists of two parts, each of which being a block of time half of total time duration.

First part: what will be completed in this sprint?

In this part, the development team works to make a prediction of the functionality that will be developed during the sprint. The product owner is the product backlog prioritized the development team and among all collaborate to achieve a understanding of the corresponding to the sprint work. The entrance to this meeting consists of the product backlog, the latest increase of product, the estimated capacity of the development for the sprint team, and performance gone from the development team. The number of tasks of the product backlog selected for the sprint it depends only on the development team. Only the development team researchers can assess what is able to carry out during the sprint starts.

Once the development team considers what tasks the product backlog will deliver in the sprint, Scrum team produces a sprint goal. The sprint goal is a goal that will be achieved during the sprint through the implementation of the product backlog, and provides guidance for the development team about why they are building the increase.

Second part: how would be achieved to complete selected work?

Once you have selected the work for sprint, the development team decides how It will build this functionality during the sprint, to form a product increase.

The selected product backlog items for this sprint, most plan to deliver them, called sprint backlog. The development team usually begins by designing the software and work needed to become a functional product increase the product stack. The work can be of variable size or estimated effort. However, during sprint planning, planning for enough work as so the development team can make a prediction of what believes that you can complete in the sprint that starts. For the end of this meeting, the work planned by the development team for the first days of the sprint will have been broken down into units a day or less. The development team is self-organized to take the sprint backlog work, both during sprint planning as according to go being necessary during the sprint.

The product owner may be present during the second part of the meeting, for clarify the selected tasks, and for helping to establish compromises. The development team determines that he has too much work or you don't have enough work, it could renegotiate the tasks with the product owner. The development team could also invite other persons to attend in order to provide advice technical or related to the domain.

At the end of the sprint planning, the development team should be able to explain to the product owner and Scrum master how to work as a team organized to achieve the sprint goal and create the increase in software.

Sprint review

At the end of the sprint a sprint review, is conducted to inspect the increase software and adapt the product backlog if necessary. During the sprint review, the Scrum team and stakeholders (customers, managers...) collaborate on what it has made during the sprint. Based on that, and any change in the product backlog made during the sprint, the assistants collaborate to determine the following things they could be made. It's an informal meeting, and the presentation of the increase has objective to facilitate the feedback and fostering collaboration.

It's a meeting restricted to four hours for sprints or block of a month. For shorter sprints, reserves a time proportionally less. By for example, Sprint's two weeks have sprint review of maximum two hours.

The sprint review includes the following elements:

- The product owner identifies what has been done and what not.
- The development team talks about what went well during the sprint, which problems appeared, and how those problems were resolved.
- The development team shows the work that has done and answers questions about the increase.
- The product owner talks about the product backlog in the current state. Esteem likely completion dates in time based on the retrieved progress to date.
- The group collaborates fully about what to do in the next place, so that the sprint review provide information valuable input for sprints subsequent planning.

The result of the sprint review is a product backlog revised. It is also possible that the product backlog received a general adjustment to meet new opportunities.

Sprint retrospective

Sprint retrospective is an opportunity to inspect a Scrum team same, and create a plan of improvements that are addressed during the next sprint.

The sprint retrospective takes place after the sprint review and before the next sprint planning. It's a meeting restricted to a three-hour time block for sprints a month. For shorter sprints it reserves a time proportionally minor.

The purpose of the sprint retrospective is:

- Inspect how was the last sprint in terms of people, relationships, processes and tools.
- Identify and sort the items most important to that they were well, and possible improvements.
- Create a plan to implement improvements to the way in which the team Scrum performs its work.

The Scrum master encourages the team to improve, within the framework of the Scrum process, its development process and practices to make them more effective and enjoyable for the following sprint. During each sprint retrospective Scrum team plans ways of increase the quality of the product through the adaptation of the definition of done according to be suitable.

By the end of the sprint retrospective, the Scrum team should have identified improvements that you will implement in the next sprint. The fact implement these improvements in the following sprint is the subsequent adaptation to the inspection of the team of development itself. Although improvements can be implemented in any time, the sprint retrospective offers an event dedicated to this purpose, focused on the inspection and adaptation.

CONFIGURATION OF SCRUM

In order to follow an implementation of comfortable Scrum for the development team, it has been decided to adopt the following configuration:

- Regarding the items or tasks, it has been agreed to define items as smaller possible, so that if an item is large is defined in two different tasks.
- Has decided to include the phase of verification of the items as requirement for consider a made item.

- The team has decided to make two-week sprints. This is a parameter that although you should not change frequently could be subject to a review after an analysis of the first sprints through the monitoring.
- The team has decided to use a physical Kanban as a support Board for the realization of the meetings during the sprint. As you can see in Table 3.1, the Board consists of five columns where the State is reflected in the items (tasks, development of functionalities, validations, etc.) to be carried out by the team. The first "Product Backlog" column contains all the items to make.

The second "Sprint Backlog" column contains the items to be carried out during the Sprint in course. The third column "In progress" contains items that are performing at this time. The fourth column "Review pending" contains the items that are already made but require a validation on the part of a Member of the team different from that carried the item. The fifth column "Done" contains items that are made and validated, according to the Definition of Done.

Table 3.1: Details of the configuration of the Kanban board

Product Backlog	Sprint Backlog	In progress	Review pending	Done

The Kanban board can be changed only during the daily meetings, with the exception of add new tasks at any time. In this way, the modifications are made in front of all the members of the team, thus facilitating everyone knows the State of the sprint.

Since the members of the team have flexible schedule, has decided to make the daily meetings every 2 days, and do it in front of the office physical Kanban board. Every fifteen days (since two weeks sprints are made) is the sprint planning to continuation of the sprint review. In no case are treated as a single meeting but entities of both meetings, compliance with their objectives, etc. The team has decided to do so for convenience of the application of Scrum, since it is the least disruptive to the development.

With respect to versioning software, working with a main branch of development during a sprint. Every developer works in an own branch, and when this implements a functionality joins the main branch of development. At the end of the features of the sprint backlog that main branch just creating a new version stable for the software that goes into maintenance.

Using the Jira (www.atlassian.com/project-mgmt) project management that conforms to the Scrum framework using a plugin called Jira Agile, this It allows that the sprints will be reflected and provides a Kanban board which is a copy virtual physical Board. It also serves to go by calculating the burndown chart in a way automatic, thus facilitating the management of development and therefore the follow-up.

CHAPTER 4

TECHNOLOGICAL BACKGROUND OF CONTINUOUS DELIVERY IN CLOUD COMPUTING

Ivan Stanimirović

This chapter will introduce both technological concepts and the necessary tools for the complete understanding of the part more technical project. This will explain the concept of cloud computing and continuous delivery. The tools that support the set of stages of the process of development or pipeline development.

CLOUD COMPUTING

Definition

Cloud Computing [1] is a model that allows on-demand access to a set shared computing resources such as servers, networks, storage, applications or services. These resources can be quickly provision of a minimum effort of administration or interaction with the service provider [2]. This cloud model mainly promotes availability [3] and is composed of five essential characteristics, three models of service and four deployment models.

Features

The main features of cloud computing are [4]:

Self-service on demand. A user can provision computing capabilities, such as server time and network storage, to the extent that you need them without need for human interaction on the part of the service provider.

Access from the network. Capabilities are available from the network and accessed using an Internet browser. For example, the cloud storage service that have documents stored in the cloud.

Multitenancy. The provider's computing resources are enabled to serve multi-tenant model, re-allocating both physical and virtual devices assigned according to the requirements of the consumers. The consumer does not have control or knowledge about the exact location of the resources that you are being provided although you may be able to specify location to data center level.

Elasticity. Resources can be provisioned elastically, in some cases automatically, to increase rapidly and can also be released elastically. For the user, these capabilities are often available for provisioning appear as unlimited and can be purchased in any quantity in any moment. For example, a scenario of sports statistical calculations has a peak of processing over the weekend, resources can be released during the week and consumption.

Monitoring. Cloud systems automatically control and optimize the use of resources through a measurement appropriate to the type of service. For example, a service that offer virtual machines monitors the number of operating hours of each virtual instance.

Service Models

There are three models of service depending on what they offer and the user they go to intended. The layout of some different service models contains Infrastructure as a Service (IaaS), Platform as a Service (PaaS) and Software as a Service (SaaS).

SaaS - Software as a Service. In this model, the user uses the provider that are hosted in a cloud infrastructure. For example, the email such as Yahoo. The user neither administers nor controls the infrastructure that supports these services, but if some configuration parameters such as the size of the input tray.

PaaS - Platform as a Service. This model allows the deployment of provider applications created by the user, using vendor tools. By example the Google App Engine platform offers a service based on this model to applications written in Java, Python, Go or PHP. The user does not control the infrastructure that supports these services. But it controls the applications or services deployed and some environmental variables that could cause, the provisioning of a new server to provide better response time.

IaaS - Infrastructure as a Service. This model allows the user to provision resources computing such as storage, processing, networks and other elements where consumers can deploy and run the software they want. An example is Amazon Web Services that even offers a catalog of systems preconfigured operating systems to provision virtual machines.

Deployment Models

Private cloud. The infrastructure of this cloud is dedicated exclusively to an organization. It can be administered by the organization or by a third party and can be within the organization, “on premises” or off, “off premises”.

Community cloud. The infrastructure of this cloud is shared by several organizations, which are normally academic in character and supports the concerns of a community on a specific topic. For example, Samsung’s cloud in collaboration with the University of Vienna, which takes advantage of the current smartphones overnight. The infrastructure can be managed by the organization or by a third party.

Public cloud. The infrastructure of this cloud is available to the general public and this infrastructure is provided by a vendor who sells services in the cloud. Cloud hybrid. It is the composition of two or more clouds, for example private and public, that remain unique entities and csoexist for having technology that allows to share data or applications between them. It is useful for example in a scenario where an application is developed and tested in a private cloud and then deployed in a public cloud.

Continuous delivery

PREVIOUS CONCEPTS

The continuous delivery [13] is the aggregation of different practices obtaining the goodness of each. Therefore, before introducing this concept researchers will define the practices that groups.

Automated testing

It consists of automating to the maximum the software testing process. The tests contribute value because they guarantee a certain level of quality of the software, but if they are also executed automatically this value is maximized. To carry out the automated testing uses specific testing software to control the execution of the tests and compare the performance result with predictable results. You can automate repetitive test tasks or you can perform test manually would be difficult to perform.

Its main advantages are:

- It saves time and money. Once a test has been programmed, it is executed automatically without the need for human intervention.
- The accuracy of the test is improved. The tests are always executed in the same way form and under the same parameters, so that you cannot forget to execute a test or overlook some error in the results. This could happen in case of manual testing.
- The coverage of the tests is increased. It has the capacity to execute a large amount of test each time. In addition, the testers by releasing themselves from running a large amount of test manually, they can dedicate themselves to programming new tests or to carry out tests on more complex functionalities.

Continuous integration

Continuous integration is a practice that consists of mixing several times a day copies of code in which developers are working towards a repository principal. This practice aims to prevent problems of integration. The most practice with automated testing, it is also a developer's job breaks the work of others. The principles of continuous integration are:

Maintain a single main repository of code.

- Each commit to the main repository must be compiled.
- Automate code compilation.
- Automate the execution of the tests for each new compilation.
- All developers must integrate their code into the main repository at daily.
- The tests are run in an environment that is a clone of the production environment.
- It should be easy to get the latest version of the executables.
- The development team should be able to know the result of the last compilation
- The advantages of continuous integration are:
- When the tests fail or a bug is found the developers can return to the latest version of the code without bugs easily. Thus there is no need to lose time looking for the source of the bug.
- Developers continuously detect and fix integration problems, avoiding problems of integration at the last moment
- Early detection of incompatible or broken code.
- Early detection of conflicting changes.
- All changes are tested (if automated testing is performed).
- A compiled version for testing, demo or release is always available.
- Frequently uploading code to the repository often causes devel-

opers to work in a more modular way, reducing complexities.

The disadvantages of continuous integration are that a start-up is required initial. And it requires a well-designed test suite to get the benefits of automated testing.

Automatic deployment

This practice consists in automating the process of deployment or distribution of the software to an environment, in the case of Ludium this environment is that of testing. Researchers must take into account that such deployment may involve the management of a service or the updating of application-dependent systems such as databases, monitoring the application or business intelligence systems.

Assuming that the testing environment should be as close as possible to the production environment, at the time of automatic deployment with this means that the potential problems of advance of the software, so that it is prepared at all times to deploy in production in a fast way.

Continuous delivery

It is a practice used in software development to automate and improve the process of delivery or commissioning of the software. Techniques such as automated testing, continuous integration, or automatic deployment. This allows software, package and deploy it to the agile way. As a result, speed and reliability are acquired in the process of improving software (either by correcting errors or by developing new functionalities) since the process is repeated frequently. This means minimizing the risks, since the process is highly known, and it also reduces the resources devoted to doing so manually.

The new software is automatically deployed to testing or pre-production environments, but never to productive environments. The reason may be due to reasons technical or business, in any case it is intended to put into production monitored version of the software. This does not mean that the production process is manual, but the deployment process in these environments should be initiated manually.

The advantages of continuous delivery are:

- Adapt to business needs. Since only the deployment is planned to production when necessary even if it is always ready.

- Adaptation to the change of development requirements. In addition, almost immediately.
- Decrease in technical debt. Technical debt is the consequence of a hasty development or sloppy deployment. And it decreases because they are making frequent deliveries.
- The technical team is released from tasks that can be automated. As consequently human resources can be dedicated to doing tasks where really contribute value to the product developed, for example the development of new tests

In short, it acquires the ability to react quickly and respond to changes, since whether they are development requirements or business needs.

Continuous deployment

The concept of continuous deployment is in essence the same as the concept of continuous deployment delivery, but the process of deploying the software in productive environments is done automatically.

An undesirable consequence may be to trigger the sufficiently tested. For this reason in this case the monitoring of metrics of the application in production is critical. This way you can detect malfunctions of the application.

Figure 4.2 shows the steps by which both procedures pass both the Continuous delivery as the continuous deployment follow a same pipeline to exception of the transition to deployment in production.

In the case of continuous deployment this transition starts automatically. In the case of continuous delivery, human intervention is required to initiate the deployment, this is known as "push the deployment button".

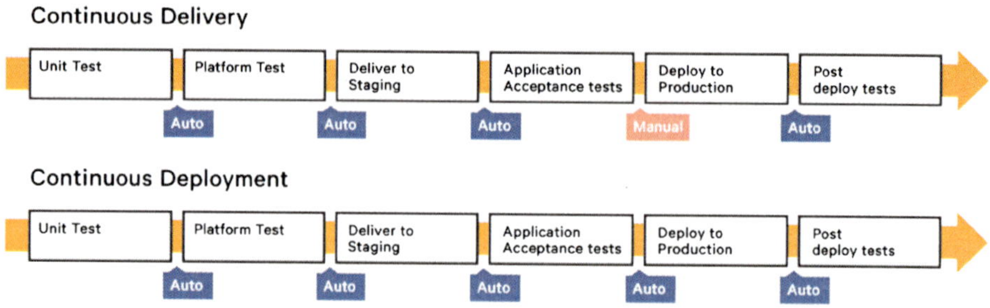

Figure 4.2: Steps by which both procedures pass both the Continuous delivery as the continuous deployment

TOOLS FOR CONTINUOUS DELIVERY

In this section will be presented the tools used to develop a pipeline of continuous delivery. It should be noted that during the development of this project the choice of open so tools is paramount. The continuous delivery, being a relatively recent concept, does not have on the market a large variety of tools of this type to choose. Of course, each of them has multiple plugins and add-ons for tool customization.

Jenkins

Jenkins [5] is an open source continuous integration tool developed in Java. This tool provides continuous integration services for the development of software. It is a system based on a server that executes a servlet container like Apache Tomcat. Supports version control tools like CVS, Subversion, Git or Mercurial. The main developer is Kohsuke Kawaguchi. Released under MIT license.

Compilations can be initiated by a new code upload to the version control, can be programmed to run every certain interval time or at the time other compilations have been completed (for example, in the case of dependencies), or through a request to a specific URL.

Puppet

Puppet [17] is an automation software tool that helps to manage the cycle of infrastructure, from provisioning and orchestration. Using Puppet you can automate repetitive tasks, perform and manage changes in a cloud infrastructure.

Puppet Performance

Puppet uses a declarative model for IT automation. Figure 4.3 shows the Puppet's life cycle, which consists of the following steps:

1. The desired state of the infrastructure configuration is defined, using the language Puppet declarative (based on Ruby).
2. Configuration changes are simulated prior to application
3. The changes in the infrastructure are deployed, correcting possible deviations from the configurations.
4. The changes between the current state and the desired state are reported, as well as of the actions to take to move from one state to another.

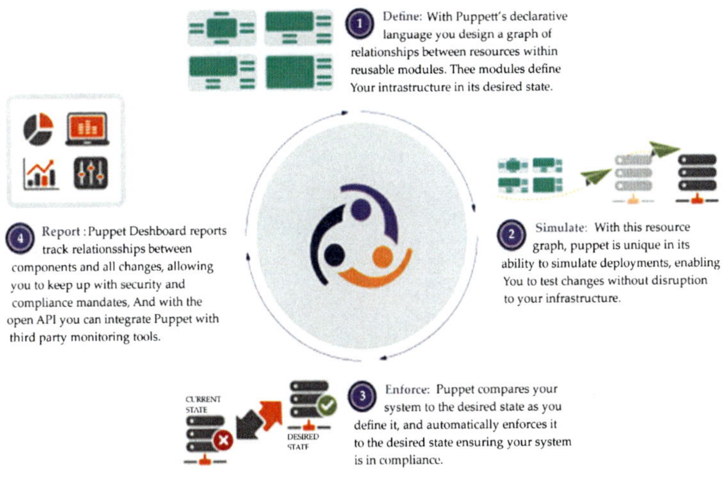

Figure 4.3: Puppet's life cycle.

CHAPTER 5

DESIGN AND IMPLEMENTATION OF CONTINUOUS DELIVERY IN CLOUD COMPUTING

Ivan Stanimirović

Prior to the implementation of the requirements specified in previous chapter, it is necessary to design and implement a development pipeline. Likewise, researchers will address the implementation of the continuous delivery and, finally, the implementation of the orchestration of the cloud.

CONTINUOUS DELIVERY

As explained in Chapter 4, the continuous delivery [17], the process of deploying continuously software versions, is based on the concept of continuous integration. It needs to be producing new versions of software frequently. TO Practice, each new integration acts as a trigger for the deployment process. By for this reason, it is necessary to design and implement a previous development pipeline to the adaptation of a model based on continuous delivery.

DEVELOPMENT PIPELINE

The development pipeline describes the software development process [14]. The figure 5.1 illustrates the design of the pipeline, starting with version control software (Source code management) used during the implementation, through the server of continuous integration (continuous integration) that checks and compiles the code and, for finally, the automation software (continuous delivery) carried out by platform deployments.

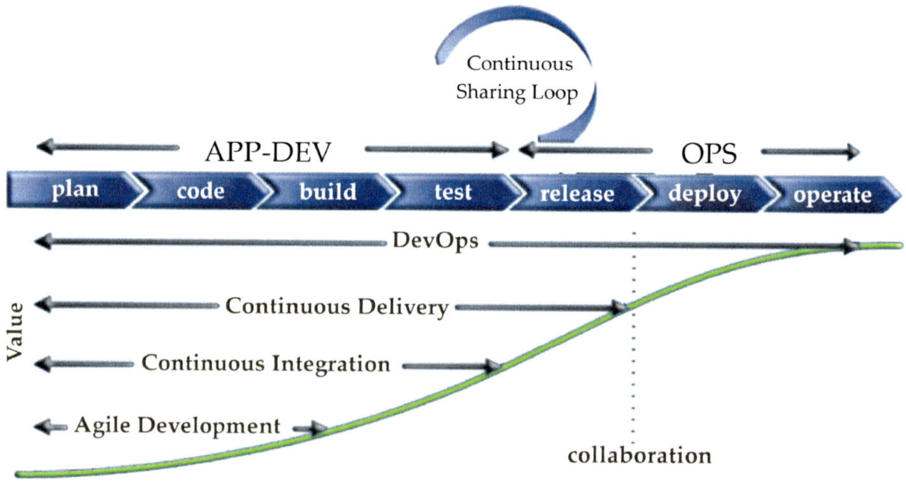

Figure 5.1: The design of the pipeline, starting with version control software

Developers send the code to the version control system via the same development framework. In this case Visual Studio 2010, which allows the integration with the version control system.

In the final implementation of the pipeline Git has been used as control software for versions, Jenkins performs the functions of continuous integration and Puppet implements the deployments automation system, all these tools have been introduced in Chapter 4.

With respect to the pipeline flow, the result of adding to the pipeline a software repository and the necessary actors can be illustrated as:

1. The developer writes a new code, which contains a new functionality or the correction of an error, and subsequently updates the Git repository.
2. The integration server continuously checks the code and compiles it.

3.a In case the compilation successfully completes the new version of the software is stored in a software repository.

3.b In case the compilation fails, Jenkins notifies the developer of such an event indicating the details of the fault.

4. Puppet monitors the software repository for new versions to deploy.

5.a In case of finding a new non-stable version of the software, it is deployed automatically to the test environment. Note that in this case the deployment is started by the same system, so a policy of continuous deployment is being implemented.

5.b If a new stable version of the software is found, it is deployed to the production environment which provides the cloud service. Note that in this case the deployment is manually initiated by the cloud-admin, so a delivery policy follows the deployment policy described in Chapter 4.

5.c In case of failure during any of the deployments the cloud-admin will be notified through the fault detection mechanisms. Then, manually start an automatic deployment of the latest version running in the environment where the deployment failed, this is done to have always running the platform in any of the environments contemplated in the pipeline. Finally, you will review the process for detecting and correcting errors thus restoring the flow of the pipeline.

Let's look at each of the components of the pipeline in detail.

Version control system

For each project a branch structure is created, this structure is represented with a diagram. A branch is a duplicate of an object, usually another branch that is under version control. The branch structure is composed of a branch development that contains the latest integration of all the functionalities developed so far, the branch DEV is the main branch. A branch is created development for each Scrum Sprint that will eventually lead to a new stable version of the software.

Before starting to work on a user history the developer creates a branch on which the development will take place, for example the DEV-401 branch. Completion of development of the user history, the branch code will be integrated into the main branch.

If researchers navigate through the file system representing the branch structure researchers will find a folder structure with a folder for each project, folder DEV. This folder contains the main branch of the project and a folder for development of each user history, the FEATURES folder.

In each integration, a software knowledge database is generated, called KDB. This database consists of the specification of the functionalities, the comments that the developers include on the development and also the detail of the files that have been affected.

Continuous Integration Server

Each time a developer adds the code of his branch to the main branch of code integration is performed. The purpose of this process is to build a new version of the software and also get different code information.

When a compilation is successful, the components are recompiled software that depend on the newly recompiled component. This information will be provides Jenkins [12] during the configuration of the integrations. Jenkins is able to graph the dependencies between software components to facilitate the maintenance of these dependencies.

Automation software for deployment

The automation software continuously tracks the installers' repository of the platform in search of a new version. The moment you find a version more recent than the one deployed in the development environment, it is when deployment of this new stable version of the software on the development platform to perform the functional test phase and user experience.

Puppet makes changes in the resources of the Servers after the deployment of a new version of the platform. It generates diverse information, the Node column refers to a particular node, researchers see an arrow at the side of the name that tells us if the deployment was successful. Latest report tells us the time stamp of the information shown. Total indicates the number of resources associated with the node. Changed is the number of resources that have undergone changes and unchanged the number of resources that have not undergone changes.

The versioning of the software is a practice that consists of the assignment of a number unique to a software to indicate its level of development, this number is increasing as the software development increases. This versioning allows to apply a policy of continuous deployment against the development environment, that is to say, insofar as it is available a new non-stable version is deployed. It also allows you to apply a policy of continuous delivery against the production environment. That is, the deployment of stable versions of the software is started manually.

The platform is versioned with four figures, the first figure indicates the project, the second increases in each sprint and is odd if it corresponds to the version in development and is even if it corresponds to the stable version result of a sprint, the third digit is the version number within the sprint and the last exactly corresponds to the Jenkins integration number. The first three figures is updated by the project manager and the fourth figure automatically maintains it the integration server continues.

IMPLEMENTATION OF CONTINUOUS DELIVERY

The process of continuous delivery is managed by Puppet that allows to define the state in which the servers must be. For this, a Puppet system will be implemented in its version 3.5.1 and will also develop the necessary status definition codes.

Components of the Puppet System

Figure 5.2 shows the components of the Puppet system and their relationships. Researchers can see that Puppet needs a central server or Puppet master, a database or PuppetDB, a management console or Puppet dashboard and, finally, the clients or Puppet agents.

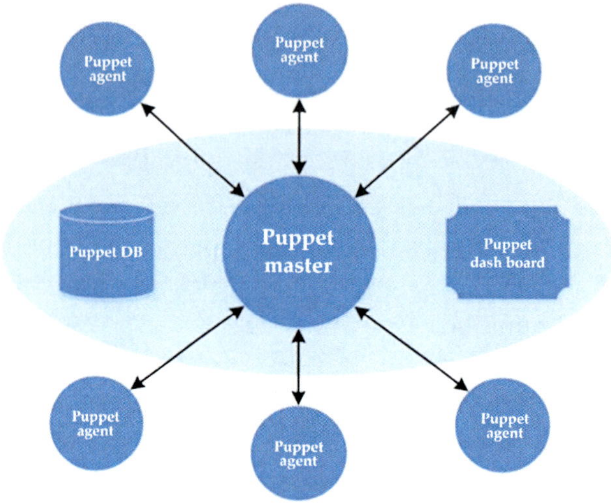

Figure 5.2: The components of the Puppet system and their relationships

Puppet master

The Puppet master or master is the central Puppet server containing the modules necessary to generate the definitions of state and is responsible for distributing them to the agents. It also keeps a record of agent connection requests, which are encrypted, and a certain certificate management is required.

As for the scalability of the master, it is capable of supporting up to 100 agents if it uses the database (HSQLDB) that integrates. For systems with more agents you can connect to a PostgreSQL cluster to scale the number of agents supported.

Puppet agent

The Puppet agent is the Puppet client that is installed on each node. Is responsible to initiate communication with the master for the application of the state definitions or cCatalogs. It also keeps the facts updated by the master. These facts are system, for example IP, if it is a virtual node or the node. The output on the agent console after the application of a catalog shows the work performed by the agent in order to apply the catalog, it also shows how the agent searches the catalog for possible future uses and the time spent in the application of the catalog. This last metric can be useful to detect abnormal behaviors.

PuppetDB

It is the database where the data are stored on the systems containing the agents, as well as facts. A REST API is available for integration with third party software or performing status reports of systems. Being a critical component in the architecture, it has a console to monitor your status, as well as other metrics and data.

Puppet Dashboard

The Puppet Dashboard is the administration console of the Puppet master, both can check the application status of the catalogs such as adding nodes, classes or grouping the nodes as appropriate. Allows you to have an overview of the Puppet system and detects a failure to show the specific error of the node in question. However, there are failures in the application of the catalogs. Each bar represents a day, the green part is the number of correct catalog applications and the red part to the applications. It can be seen that there is a day when there is practically no have applied the catalogs, this can be due to a real failure in the application or to that it is the dashboard itself was not in service that day and therefore could not record application reports.

Now researchers know the main components of the Puppet system. Agents are connected bidirectionally with the master since they report the facts and receive from the master the catalogs to apply. They are also the agents that report the facts to the PuppetDB through the master. And by lastly, the Puppet dashboard connects to the master and the PuppetDB.

Puppet manifests

The way that Puppet has to define the state of the machines is through the manifests. These are files written in Puppet code that is based on Ruby.

The language used is Puppet's own, it is declarative and for each resource it allows to describe its characteristics. By convention these characteristics are encapsulated in classes and adding classes are generated modules. Modules are associated with the nodes or group of nodes (all nodes in the group are treated identically) by means of manifests special: site, nodes and modules.

The classes use the existing facts in the Puppet system to modify their behavior. For example, depending on the operating system of the node, a configuration can be found in different directories. It is also possible to define new facts that happen to be available in the classes for your consultation.

For the implementation of the requirements specified in the previous section, defined a new fact in the implemented Puppet system. It is a fact that informs of the version of the platform installed on a particular node.

Puppet Performance

Figure 5.3 illustrates the data flow between the Puppet components. At all times the data travels through a secure channel, since its interception could potentially be reviewed.

Once configured, the agent collects information about the node and sends it to the master (step 1). The master compiles the definition of the state in which it should find the node, this definition is called a catalog and researchers will see how you get more forward (step 2). At that time, the agent makes the necessary changes at the node to get the status defined in the catalog and informs the master and the dashboard to that the changes are visible in the management console (step 3). By last, note that the master has a public API that can be queried by software of third parties, thus facilitating integration with other systems (step 4).

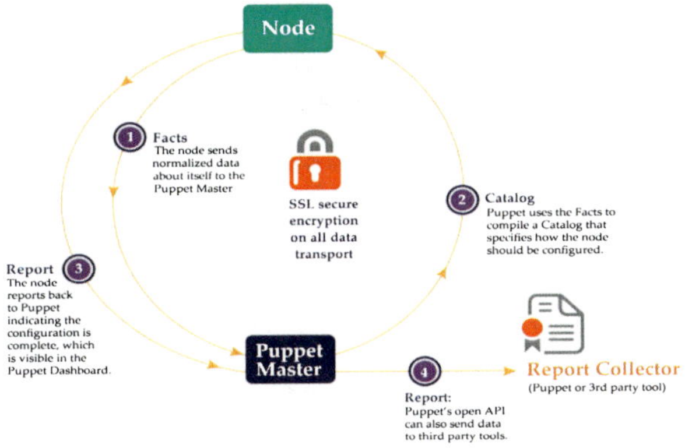

Figure 5.3: Data flow in the Puppet system.

The process of compiling catalogs or defining the state of a node is complex. This process takes into account data inputs that come from different components of the Puppet system. The compiler chooses the manifests of the modules assigned to a specific node, resolves the variables in them with the data it has received and finally compiles it in the form of a catalog, which is what is finished sending the agent.

Thus, in the case at hand the catalog compiler takes into account the site, the Dashboard or console, the information contained in the PuppetDB, the facts it provides the agent and, finally, the modules installed in the master itself. From the console or dashboard you can create new classes as well as assign them to existing agents.

The compiler receives the list of available classes defining the classes and their settings. This information is provided exclusively by the modules. The compiler treats of completing this information with the inputs of the other components, for example assigning the value of the facts to the parameters of the modules to obtain the desired behavior that will eventually be reflected in the catalog.

It is also possible to assign a value from an external resource to the PuppetDB, for example the time zone in which researchers want to operate the nodes. If researchers do it when registering a new agent in the Puppet system, irrespective of its geographic location, the time of the Node will be automatically set to the desired time zone.

The compilation of the catalog in the master occurs on demand, the agents go asking regularly the catalog of changes to the master. Figure 5.4 shows the data flow between the agent and the master.

The agent initiates the request from the catalog to the master, sending the name and node. This is when the master checks the classes it has to compile in function of the request data and compiles a catalog that sends the agent. The agent initiates the implementation of the catalog by consulting the current state of the resources affected by the catalog, for resources that are in a different state than necessary changes. This is where the potential of Puppet is found, without the Cloudadmin has to specify the actions required to achieve the desired state, Puppet manages to reach it. This is because it is able to translate the state definition to concrete actions. Finally, the defined state is reached and the master of such event.

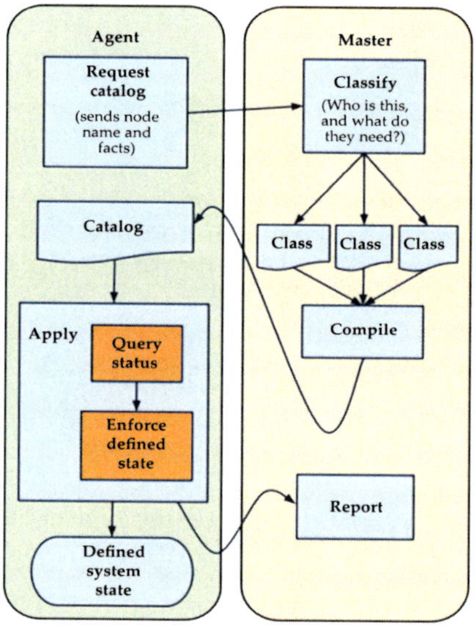

Figure 5.4: The data flow between the agent and the master.

IMPLEMENTATION

The implementation of each of the described in the specification. To implement this objective researchers must be able to provision new application in an automated way. This can be implemented in two ways: through a disk template or creating a Puppet module to get the status wanted.

CHAPTER 6

VALIDATION OF CONTINUOUS DELIVERY IN CLOUD COMPUTING IN THE REAL ENVIRONMENT

Ivan Stanimirović

This chapter introduces the real environment in which the implementation of the method previously described is being validated.

INTRODUCTION TO THE ENVIRONMENT

For the validation of the implemented use cases an environment has been configured with the same characteristics as a real environment.

In total researchers have 5 virtual nodes and a private network that communicates them. Three nodes correspond to thc Puppet system (one master, one puppetdb and one dashboard) and two of the nodes will act as cloud applications with Windows operating system, where researchers will install the agents. The topology is the same as shown in the previous chapter.

The different tests were carried out after setting up the Puppet system. In this configuration the master uses the puppetdb as the data store and the dashboard has access permissions to consult.

VALIDATION

Below researchers will describe the tests performed with the objective of validating the correct implementation of use cases.

Application in cloud scaling

Purpose of the test

Validate that the use case of the cloud scaling is implemented following the specification. That is, it is possible to provision new nodes and add them to the cloud of applications when more users are needed.

Steps to follow

- Researchers start from a node in which the platform is not installed.
- Install the Puppet agent and registere the node in the master assigning the module.
- The platform and its dependencies are automatically deployed.
- Manually add the node to the application cloud.

Parameters to be analyzed

It will be verified that, in fact, the platform is installed in the version and is in execution.

Generation of errors

Error generation has been forced by disconnecting the master to verify that the agents are able to apply the last received catalog, which they cache, in case of not receiving a catalog by the master.

Analysis

Researchers will use the Puppet Dashboard, the interface of the console provided by Puppet, to follow the correct operation of the tests.

In case the master is not available, the nodes would not be able to receive a new catalog with the changes to apply. The first error message often says Connection refused. Researchers can see that the agent is still running, even if he does not receive the catalog, trying to send the facts to the master. Finally, since it had not received a new catalog from the master, it applies the last catalog that received, that is curled in the node.

Once the agent configuration problems were solved, the platform launches successfully at the node.

Upgrading the in-service platform

Purpose of the test

Validate that the use case of the in-service platform upgrade is implemented following the specification. That is, at the moment in which it is found developed a new version of the platform will be deployed in production.

Steps to follow

- Manually indicates in the module the new version of the platform that must deployed on servers in production.
- Automatically the agents of the nodes reinstall the platform to the new one version and put it into service.

Parameters to be analyzed

Researchers validate in the console that the value of the fact that the current version corresponds to the version indicated in the module. Researchers can also validate it by consulting the facts of a node in concrete.

Generation of errors

Errors have been attempted by instructing Puppet to update a platform with an earlier version than the one already deployed. The result has been satisfactory, since Puppet realizes that the platform is in a later version than the one tried "Update" and does not make any changes.

Analysis

It is important to correctly configure the parameters to be included in reports. An insertion failure in the database is resulting from a mismatch in coding. After validating this objective on several occasions, error can be found. The coding of the values must correspond to the coding of the base of data where they are stored.

Note that although the output of the dashboard does not explicitly point to the source of the error, it is very useful for its diagnosis and subsequent correction. In this case it is responsibility of the cloudadmin who is the one who operates with the systems.

Mass application of a new cloud configuration

Purpose of the test

Validate that the use case of the massive application of a new cloud configuration is implemented following the specification. That is to say, that at the moment develop a new configuration to be copied to the servers and restart the service

Steps to follow

- Manually store files containing a new configuration of the cloud in the master, replacing the files of the old configuration.
- Automatically the new configuration is deployed in the application cloud agents, who also restart the service at the nodes so that apply the new settings.

Parameters to analyze

The files containing the new configuration have been deployed to the cloud applications. You can check the contents of the files in a specific node, or it can be consulted in the dashboard if there have been changes in what files are refers. The restart of a service is validating in this way the application of the new configuration.

Generation of errors

In this case the generation of errors has not been forced.

DEVELOPMENT ENVIRONMENT

This section will introduce the development environment used for the presented technique: the operating system used, the language of the programming and the applications used.

For the realization of the project a server has been used in semi-tower format with Linux, specifically Ubuntu Server 12.4 LTS. Ubuntu has been used for its performance and stability, and this particular version being the most recent version with long-term support. This server hosts the VMWare virtual machines that is used in many development environments.

The minimum requirements of the Puppet master hosted in one of the virtual machines are 2 processor cores and 1 GB of RAM. It has been provided a core and 2 GB of RAM, with this configuration hardware has remained stable at all times. For the writing of the memory has been used MacOSX.

The programming language has been Puppet's own, which is based on Ruby. Ruby has also been used for the contribution to Puppet's Facter module explained in Chapter 2.

The applications used during the realization of the project have been Geppetto, Microsoft Office, Adobe Photoshop and ArgoUML.

Geppetto is an IDE based on the recently acquired Eclipse development environment and supported by PuppetLabs. It was created by the company Cloudsmith in form of plugin for Eclipse, and is currently as an IDE result of an Eclipse fork. In the version 4.1, the great advantage that it has is that it allows queries in real time to the PuppetDB. As a disadvantage to mention that being a very important tool specific and recently created is very little documentation on its use.

Microsoft Office is an Office suite developed by Microsoft. In its version 14.4, in particular RESEARCHER used Word for writing the text and Visio for the realization of the figures of this memory. They have the advantage that the two are perfectly integrated programs, as they belong to the same suite. The disadvantage is that since it is a payment suite you have to acquire a license, but fortunately RESEARCHER have access to one belong to the program Microsoft Bizpark. Adobe Photoshop has been used in version 5.2. This image editing program is very complete and allows, among other things, to cut the exposed screenshots in the memory and also the scaling of the images to adapt them to the A4 size of the memory. The great advantage is that it has a great diversity of functions. As a disadvantage to mention that its management can become quite complex.

ArgoUML is a UML diagramming application written in Java, developed by CollabNet Inc. and published under the BSD License. The advantage it brings is its ease of use due to the simplicity of its graphical interface. The disadvantage is that it is little stable.

VMWare Workstation is the machine virtualization application used in this book. A temporary license has been used. The advantage it brings is the great flexible configuration of private networks between virtual machines, with support for advanced settings.

VARIATIONS ON INITIAL PLANNING

The project has undergone variations with respect to its initial planning due to some complications arising during implementation and tasks not included in the initial planning.

The platform changed to half a project to a version with a radically architecture different, causing to have to redo the implementation of the specification of the Chapter 5. Although this complication has not added much difficulty to the project, since researchers have been an active part in the architecture design of the new version, has contributed to the delay in the completion of the project.

Develop the Puppet module for a version of Puppet not stable, since it had greater support for Windows-based agents. The origin of this complication has the lack of documentation, although it is necessary to emphasize the range of new features offered by the non-stable version of Puppet. Currently, the version that works, which includes most of these functionalities, is already stable.

Tasks not included in the initial planning are: correction of a bug in Puppet's facter module for Windows-based agents, resulting in a pull-request to Puppet's master repository. This task has been discussed in Chapter 3.

CHAPTER 7

ADAPTIVE DEVELOPMENT PROCESSES FOR AGILE SOFTWARE ENGINEERING: METHODOLOGY AND PRACTICE

Ivan Stanimirović

INTRODUCTION

Worldwide budgets are raised for software development. Much of these budgets are spent on gaps, cancellation of projects and software maintenance. Poorly implemented requirements, delays of schedules, poor product quality and difficult software maintenance are recurrent problems in development projects. Among other causes, is due to poor or no administration of the development process, leading to ignore the characteristics of the project when choosing the best practices that will lead to cable at each stage. These characteristics include, for example, development and human resources.

Some answers to these problems are initiative of the area of Software Processes. It's like this as have emerged proposals such as the CMM, CMMI, SPICE, and TSP among others.

These proposals have in common the fundamental idea of Continuous Improvement. Nevertheless, certain types of organizations and certain types of projects, it is not feasible to implement these because of their complexity, the number of suggested practices, the rigidity of requirements, accompanying costs, etc.

On the other hand, lightweight process proposals such as those presented in [2], [10], [13], among others, emphasize the adaptability of the process. Generally, principles that sometimes include paperless practices and / or metrics. Since a process is not expected to be predictable, more work is being done on the preparation to adapt to change, than in the opportunity to capitalize on what has been learned so that the impact of change is minimal.

Whatever the defined process initiative that a development team natural, an organization gradually matures to the rhythm of its own needs. The process maturation includes the inclusion of practices in the development lifecycle. If the practices are effective and the processes are useful, the teams institutionalize and improve them.

In this article researchers present QualDev Process, a proposal of processes of development of software for projects that researchers call agile. The characteristics of these projects are mainly the small size of the development groups, not much experience with the technology, high staff turnover and geographic dispersion of all or part of the working group.

Our proposal is also based on the idea of continuous improvement. Do not predefine a strict order about the processes, nor does it expect the result of them to be completely predictable On the contrary, as the group or organization matures, they leave.

Adapting processes and generating ability to respond to change. The improvement continuous, although it consists concretely in the adaptability of processes (cardinal principle of lightweight process), is based on the administration of development metrics that allow the learning and maturing processes.

The organization of the article is as follows. In the second part researchers present a description of the context of the projects and groups that researchers want to support. In the third part, researchers present the framework of processes, principles, organization of processes suggested, and the methodology for the documentation of these. In the same way, researchers discussed briefly some points of implementation of the proposed processes (how to do it). By last, some results achieved until now with a particular development group, conclusions and future work.

CONTEXT OF AGILE PROJECTS

In principle researchers consider that it is not possible to expect to have a single model of development suitable for all development contexts. Software projects may present among its characteristics, non-homogeneous work teams, high staff turnover, little availability of resources (geographical dispersion), new technology, variable size of the development team, difference between the sizes and nature of the software to be developed or to maintain, among others.

The characteristics of the agile projects correspond to companies of software development youth, or groups that conform to develop a specific project. By companies young people researchers refer to companies with less than a year of permanence in the industry, with less than 2 development projects, with work in leading technologies and processes immature software development.

In the case of agile projects, concrete practices are required that in other contexts may not be as important. Examples are given below.

Experience in development technology

Given the low technology experience, the teams require more effort in the area of investigation and development. The results should be reflected in prototypes or spike solutions that support learning in technology. In parallel, it is necessary to invest in the process of knowledge management, so that the learning curve of the whole equipment is minimized.

Experience in business logic

Entering the new business market is fundamental to the economic growth of a development team with agile projects. Loss of development opportunities is critical when a development team is starting to enter the market. When you do not have experience in the business logic for which it is developed, it is necessary to emphasize the validation of the requirements analysis and in general in the validation tests. This requires a much closer contact with the customer. If, on the contrary, there is ample experience in business logic, customer contact may be less, minimizing the risk of an inadequate specification of requirements.

Experience in processes and knowledge of development metrics

The experience to follow and adapt development processes within a project is fundamental to maximize the productivity of a computer. Leading a process of understanding and adapting a life cycle model of development is indispensable within a team without much experience. For the same reason, the non-existence of historical data leads to poor resulting in delays within projects. In the context of the defined development, the team must be willing to rapidly obtain data that will enable better estimate. If the experience in development technology is high, it does not imply that formal historical data. On the other hand, if you do not have experience, the absence of these data.

Rotation of equipment

The high rotation of the equipment can be given at various levels. Companies with development projects big losers many of their developers in the course of their projects. Companies with small developments lose part of their developers at the end of their projects. This implies the search for new developers every time a project is started. No matter what the cause, the rotation of the equipment is a determining factor for the success of a development. Teams that have this feature must implement appropriate processes knowledge management, and in general, maintenance of documentation to support the continuous loss of its members.

Communication of the equipment (physical dispersion of its members)

The communication of a development team directly influences the success of the projects. Before poor communication or synchronization, the team may find itself facing a crisis of individual and individual work by each member, without achieving a group unity to support the work of achieving a final common goal. A development team without conditions communication must implement processes that change this condition. The configuration management and monitoring of the documentation and communication process are the processes responsible for working in this area.

As you can see, practices that may be unnecessary in their rigor, are made in a context defined as that of agile projects. In an agile project, many it is sometimes believed that the software process, rather than being a development support tool, is an extra burden for participants. This may be true if the approach to include it is prescriptive and includes many practices at a time. This is only achieved by leaving the practices sideways quickly as its usefulness is not seen and developers feel that by following the process, do not develop the product.

Researchers believe that even in difficult implementation contexts such as projects agile, the task can be achieved if an adaptive approach is used and the same gradually make adjustments and more practice as mature systematically include process practices from which no benefit will be obtained.

Short or medium term, imposes resistance within the group. Practices should be included in as the group discusses them and considers that they offer added value. Thus, after the inclusion of each practice, this is adapted by the team to the extent of the need seen.

Each practice can mature independently of the others. Unlike proposals like that presented by eXtreme Programming (XP) [3], researchers believe that certain process practices are not suitable for all contexts, even after they are adapted. An example might be programming in pairs or working in groups. Within the context of agile projects the developers are dispersed most of the time so that follow-up meetings with the physical presence of all the team members is already an important achievement. For this particular case, you should seek practices that support the quality of the product, without imposing conditions that nature of the team are difficult to achieve. Processes such as TSP [5] are intended to be used in any context of development. According to their proposal, practices should be included regardless of the team's criteria. Although they later promote their adaptation, some teams do not require certain practices for their development context. Worse still, some teams that do impossible to include them according to the specification of the proposal due to its complexity and rigor.

A point where emphasis should be placed since it applies in any context, it is the importance of use tools to support the process. Whatever the process practice included, it should seek tools to support the follow-up of this practice, help thus promote the continuous improvement of the practice while downloading manual operative work to the members of the team.

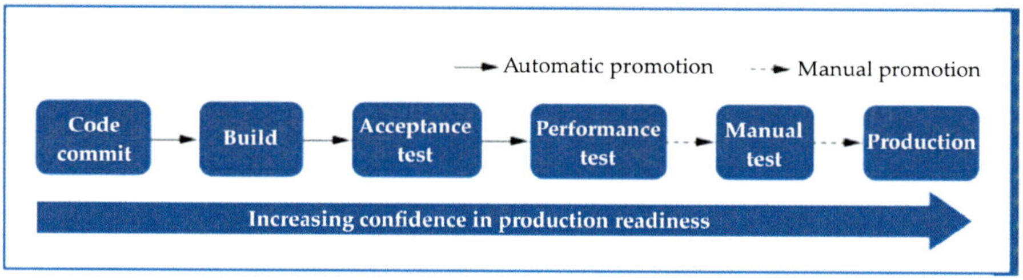

Figure 7.1: Continuous data pipeline principles

These ideas have been included in the framework of processes that researchers have called QualDev and that presented in the next section.

QUALDEV PROCESS

QualDev Process is a frame of reference for development processes for agile projects in where the project is developed in iterative cycles in which the product is constructed incrementally. QualDev process is composed of several elements. The first element is principles, the second the organization of practices, the third the methodology for implement each practice and finally a suggested implementation plan. In this section researchers study the first two elements and briefly present the proposed structure for document the implementation practices of the processes. The implementation plan is out of the scope of this section.

Principles

The QualDev software process revolves around five principles on which the practices and continuous improvement. These are:

(1) There should be planning and monitoring. Doing this practice consistently is what you will allow that the development group grows and can adapt the process to their needs. Around the principle of planning and monitoring are three more:

(2) Risk management,

(3) Quality management, and

(4) Knowledge management. The principle of administration of knowledge is what enables the results obtained with respect to the practices to be durable.

The fifth (5) principle refers to the use of tools to support the development processes.

1. Planning and follow-up.

The planning of a software project must be carried out throughout the entire project. Because different factors (internal and external), it is not possible to predict the changes that the plan will suffer.

However, planning should be done as the basis for feedback on the collection process of metrics that allow better estimations. Whatever the cycle model of plans should be continually revised in order to be adapted to the needs of the project. The follow-up to the project results in the establishment of practices that lead to adaptability and continuous improvement.

In the topic of planning a determining factor are the development cycles. These cycles should be short and iterative, and at the end of each a functional system must be completed in full. Functional prototypes that are not for specific learning inside the computer, or as user interface drafts.

It is important to define a set of objectives at the beginning of each development cycle. These should be accompanied by metrics that allow to evaluate the reach achieved at the end of the cycle. Of equal all planned activities should have definite and quantifiable commitments.

2. Risk Management

Risk is the fundamental component in defining the process that is established. At QualDev process is established that perform the basic activities of risk management, before starting the project, it is fundamental to determine the set of processes that must be followed. List the potential risks of the project, classify them, and prioritize the team to think about the best practices to minimize the occurrence of these, or the impact to an occurrence. The classification of risks should be based on a defined taxonomy, for example that presented in [4].

This principle is analogous to a risk-preventive established plans that in this case result in processes. Continuous monitoring of processes planned, introducing new practices based on the risks team finds as potential, allows processes to be introduced by self-motivation of the team before the identification of the need.

3. Knowledge management.

Knowledge is an intangible resource on which to establish metrics that allow to manage it. All knowledge brought in from outside, built, or acquired by the team must be valued as one of the most valuable assets and managed according to a strategy established. The team must manage it so that it becomes an element internal differentiator that allows to improve the productivity and quality of the work, and at the competitive advantage.

One of the most important elements that are based on the administration of knowledge is the definition of the roles within the development team. Unlike the proposed TSP [5], researchers believe that it is not enough to define roles that in general point to the part management of the project. It is necessary that clearly within the team there are leaders defined and identified by each area of knowledge, especially those that have been identified as generating the greatest number of potential risks. Each non-administrative leader must be responsible for managing a knowledge database that enables the entire team become more productive. For example, it may be necessary to define a need to modify the software testing process, or the inclusion of new tools of support to the process.

4. Quality management

The management of quality cannot fail to be one of the fundamental principles for the software development. Strategies should be defined so that throughout the development cycle can control the quality of the products being built. In no way should it be limited quality management to perform unit, functional or system tests.

Therefore, the quality strategy should be based on product inspections of work intermediaries. Although carrying out inspections on all the products of work would be ideal to ensure high levels of quality, the cost of this is so high that it would not be possible to maintain some type of competitive advantage based on the productivity of the equipment. There must be a balance in the number of inspections carried out, the number of standards and instructions which are defined, and the tests performed on the software being built.

As mentioned, the four principles presented are the basis for the definition of QualDev processes. Likewise, they are the basis for implementing a plan of implementation and adaptation to the own context of development of a work team. In addition, it is important to highlight the importance of the definition and documentation of how it should be each process / activity of the development lifecycle.

5. Support processes with tools

Software processes must be supported on tools that support their development. It is possible to carry out some processes manually without an information system that helps administration. However, some processes that are not based on a tool are impossible to carry out. Similarly, those that can be performed without tools can be better exploited if a tool is defined for its administration and management. The process support tools must be introduced incrementally, and at the initiative of the team. In the same way as the process, the tools must be adapted of development cycles to meet the needs of the team.

Organization of practices

The processes have been divided into two groups: support and development. Support processes the development throughout the whole life cycle of the construction of the product. The development refers to activities focused on the production of artifacts that are an integral part of the product.

The methodology of each practice

The practices of the support and development processes go beyond what to do and go accompanied by how to do it. We do this as implementation guides and methodology associated with the activity. The implementation guides include the inputs and outputs of each activity, along with the tools and supporting documents that should be used. The methodology includes list of tasks and order of execution of these, within each activity.

The definition and documentation of each activity is presented in the following order:

1. Process description.

A. Motivation to carry it out

2. Description of the associated activities.

A. Justification

B. Purpose

C. Implementation guides

D. Methodology (tools and supporting documents).

By way of illustration we present some aspects of the methodology for the practice of planning and monitoring. For reasons of space we cannot present it in its entirety. This practice includes a justification, a proposed process and for each activity of the process proposed a description of how to carry it out.

Justification

Given the physical dispersion characteristic of the equipment, the planning follow-up meeting is one of the few spaces that the team has to meet and feedback. Inside of meeting, the communication between all the members of the team is direct and in person. This type of communication is important for a general working team environment.

Planning review within the meeting, with input from the entire team, helps to generate an environment in which everyone has a vote in the plan that is created. Follow-up meetings allow all team members to know the summary of the presented in the development period, providing ideas to solve them. On the other hand, it allows to know the overall progress of the project, and contribute ideas to achieve the objective development.

Purpose

Planning follow-up meetings are held at the beginning of each planning period. These meetings, preferably presentations, aim to make a planning follow-up at the previous follow-up meeting. Within the meeting the activity plan is defined until the next planning follow-up meeting. Based in the activity plan, developers work by reporting their development times to keep track of the time spent in the cycle. At the time of arriving at the meeting, the time invested have been consolidated, which allows to know the general state of the activities planned at the previous meeting, and plan the activities until the next meeting.

The physical presence of all team members in the follow-up meeting is important. However, given the characteristics of agile projects, the absence of some member. The impact of this eventuality is minimized by reporting all the results of the meeting and publishing them for the knowledge of the whole team. In addition, it is expected that possible, the member of the team missing the meeting reports their absence earlier and sends a document with the points you have to exhibit at the meeting.

RESULTS AND XP METHODOLOGY SUGGESTIONS

The process presented in this chapter is being used in several software to the interior of the University. These projects the characteristics of the agile projects that we presented previously. One of these projects' purpose is to build software to support the administration of software configurations. For almost two years we have been part of the team more than 20 people with an average of 5 at the same time. The group is satisfied with students of undergraduate, master's and faculty members, none of them working full time in the project or sharing a common workspace. At the same time, software developed by the team, incorporate state-of-the-art technology with a high level of quality.

These products are developed to be put into production in the industry, achieving a transfer of knowledge at the technological level and as tools that support the process of software development.

In general, the results of experiences gathered within QualDev have been the basis to structure this proposal. Similarly, on the other hand, without the proposal of the reference would not have been possible the orderly and productive work of the team.

The acquired experiences and the software constructed, aim to be transferred to organizations with similar contexts. In this way, adaptation processes validate the development methodology. In parallel, it allows projecting the expertise gained to be useful to organizations with different development contexts.

The agile XP methodology arises with the purpose of transform the essence of the development of processes that are executed at the time of the planning and execution of a project to create this methodology focuses on integrating into continuous improvement to the user and the group of individual who will be responsible for solving perceived problem.

XP in comparison to the methodologies is faster because it entails less protocol, which avoids the existence of hierarchies within the group, which allows each member of the group can contribute in any reason, for which it is implementing or make use of this methodology is that the focus on short-term results.

The results that are obtained throughout modulation will be verified instantly and existing any anomaly or fault will be made corrections which are corresponding. This quick way is possible to obtain the expected results in the short term and efficiently, the system will become robust in the use of the modules will be achieved in a snapshot.

The results are verified by the user to approve the product and thus obtain a security and soundness in the development of the system, of this way the final result will be able to satisfy in expectations and the needs of the user.

Extreme programming or XP is a methodology of development that would be included within the known as Agile methodologies in which highest priority in achieving results and reduces the bureaucracy that uses the methodologies traditional.

Generally the development process carried associated with a marked emphasis on process control through a rigorous definition of roles, activities and artifacts, including modeling and documentation detailed. This “traditional” scheme to address the software development has proven effective and necessary in large projects (with respect to time and resources), where a high level of degree of ceremony in the process. However, this approach is not the most appropriate for many current projects where the system environment is very changing, and where it is required to reduce drastically the times of development but maintaining a high quality.

Given the difficulties in using methodologies with these time constraints and flexibility, many development teams are resigned to disregard the good practices of software, assuming the risk that this entails.

In this context agile methodologies emerge as a possible answer to fill that gap in methodological approach. By being especially geared towards small projects, Agile Methodologies constitute a tailor made solution for that environment, providing a high simplification that despite this does not give up essential practices to ensure the product quality.

XP is an agile methodology for the development of software and basically consists of adjusting strictly to a set of rules that focus on the customer's needs to achieve a good product quality in a short time, focused on enhancing the interpersonal relations as key to the success of the software development.

The philosophy of XP is to fully satisfy the customer's needs, that is why it integrates it as a part of the development team. It promotes teamwork, focusing on time of the learning of the developers and establishing a good working climate.

This type of programming is appropriate for projects with imprecise, highly changing requirements and where there is a high technical risk.

XP is designed for application development which require a small group of programmers, where communication is more feasible than in groups development. Communication is a point important and should be done between programmers, project managers and customers.

These are the values of the XP approach:

Communication

Prevails in all practices of Extreme Programming. Face-to-face communication is the best form of communication between developers and the client. Very agile method. Thanks to this, the team is able to make changes that the customer did not like.

Simplicity

Simplicity helps developers to find simpler solutions to problems, as the client stipulates. The developers also create features in the design that could help solve problems in the future.

Feedback

Continuous customer feedback allows the developers lead and direct the project in a right direction to where the customer wants.

Courage

Requires developers to go hand in hand with change, because we know that this change is inevitable, but being prepared with a methodology help that change. Program for today and not for morning.

Respect

The team must work as one, without doing decisions. Extreme Programming promotes the work of the team. Each member of the project (Client, developers, etc.) are an integral part of the team responsible for developing quality software. The team must work as one, without making decisions sudden.

Although other sources of information appear some variations and extensions of XP roles, in this we describe the roles according to the Beck's original proposal.

Programmer

The programmer writes the unit tests and produces the system code. There must be a communication and coordination between programmers and other team members.

Client

The client writes the user stories and the functional tests to validate their implementation. In addition, it assigns priority to user stories and decides which are implemented in each iteration focusing on providing greater value to the business. The client is only one within the project but may correspond to an interlocutor who is representing several people who will be affected by the system.

Test Manager (Tester)

The test taker helps the client write the functional tests. Run the tests regularly disseminates the results to the team and is responsible for test support tools.

Tracker

Tracker provides feedback to the team in the XP process. His responsibility is to verify the degree of accuracy between estimates made and the actual time spent, communicating the results to improve future estimates. It also monitors the progress of each iteration and assesses whether the objectives are achievable with time and resource constraints gifts. He determines when it is necessary to perform change to achieve the objectives of each iteration.

Coach

He is responsible for the overall process. It's necessary that know the XP process to provide team members in order to XP practices and follow the process correctly.

Consultant

He is an external member of the team with a specific knowledge in any subject necessary for the draft. Guide the team to solve a specific problem.

Manager

It is the link between clients and programmers, help that the team effectively works by creating the adequate conditions. Its essential work is coordination.

The XP methodology defines four variables for any software project: cost, time, quality and scope. In addition, it is specified that, of these four variables, only three of them may be arbitrarily fixed by external actors to the group of developers (clients and project managers). The value of the remaining variable may be established by the development team, depending on the values of the other three. This mechanism indicates that, for example, if the customer establishes the scope and quality, and the project manager the price, the group shall be free to determine the time the project will last.

For this, it is a question of realizing cycles of development short (called iterations), with deliverables functional at the end of each cycle. In each iteration, it performs a complete cycle of analysis, design, development and tests, but using a set of rules and practices that characterize XP.

Typically a project with XP takes 10 to 15 cycles or iterations. An XP project succeeds when the client selects the business value to implement based on the ability of the team to measure functionality can deliver over time. The development cycle consists (roughly) in the following steps:

The client defines the business value to be implemented.

The programmer estimates the effort required for his implementation.

The client selects what to build, according to their priorities and time constraints.

The programmer builds that business value.

Go back to step 1.

In all iterations of this cycle both the client as the programmer learn. Do not push the programmer to perform more work than the estimated that quality will be lost in the software or will not be met the deadlines. In the same way the customer has the obligation to handle the scope of delivery of the product, to ensure that the system has the highest value of possible with each iteration.

While the lifecycle of an XP project is very dynamic, can be separated into the following phases:

Exploration,

Planning of the Delivery,

Iterations,

Production,

Maintenance and

Death of the Project.

Phase I: Exploration

At this stage, clients are broadly interested to the first product delivery. At the same time, the development is familiar with the tools, technologies and practices that will be used in the project. The technology is tested and the possibilities of system architecture by building a prototype. The exploration phase takes few weeks to a few months, depending on the size and familiarity of programmers with the technology.

Phase II: Delivery Planning

In this phase the client establishes the priority of each user history, and correspondingly, the programmers make an estimate of effort necessary for each of them. Agreements are the content of the first delivery and determines a schedule in conjunction with the client. A delivery should be obtained in not more than three months. This phase lasts a few days.

Estimates of effort associated with the implementation of the stories is programmers using the dot as a measure. A point, is equivalent to an ideal week of programming. Stories are usually worth 1 to 3 points. On the other hand, the development team maintains a "speed" of development, established in points by iteration, mainly based on the sum of points corresponding to the user stories that were completed in the last iteration.

The planning can be done based on the time or reach. The speed of the project is used to establish how many stories can be implemented before a certain date or how much time will take to implement a set of stories. To the planning by time, multiply the number of iterations by the speed of the project, determining how many points can be completed. To the plan according to the scope of the system, the sum of the user stories selected between the speed of the project, obtaining the number of iterations necessary for its implementation.

The result of this phase is a Delivery Plan, or "Release Plan".

Phase III: Iterations

This phase includes several iterations on the system before being delivered. The Delivery Plan is composed of iterations of not more than three weeks.

In the first iteration one can try to establish an architecture of the system that can be used during the rest of the project. This is achieved by choosing the stories that force the creation of this architecture. However, this is not always possible since it is the client who decides what stories will be implemented in each iteration (to maximize business value). At the end of the last iteration the system will be ready to enter in production.

The elements to be taken into account during the elaboration of the Iteration Plan are:

User stories not addressed,

Speed of the project,

Acceptance tests not exceeded in the previous iteration

Tasks not completed in the previous iteration.

All the work of the iteration is expressed in programming tasks, each of which is assigned to a programmer as responsible, but carried out by pairs of programmers.

Phase IV: Production

The production phase requires evidence and performance reviews before the transferred to the customer's environment. The same time, decisions should be taken on the inclusion of new features to the current version, due to changes during this phase.

It is possible that the time taken for each iteration, from three to one week. The ideas that have been proposals and suggestions are documented for their subsequent implementation (for example, during the maintenance).

At this stage no further developments functions, but adjustment tasks may be required. ("Fine tuning").

Phase V: Maintenance

While the first version is in production, the XP project must keep the system at the same time as it develops new iterations. To do this, you need support for the client. In this way, the speed of development may go down after the system in production. The maintenance phase may require new staff within the team and changes in structure.

Phase VI: Death of the Project

It is when the client has no more stories to be included in the system. This requires that they be met customer needs in other aspects such as performance and reliability of the system. The final documentation of the system and are no longer changes in architecture. Death of the project also occurs when the system does not generate the benefits expected by the customer or when there is no budget to maintain it.

The XP methodology has an important set of rules and practices. Generally, they can be grouped in:

Rules and Practices for Planning

Rules and Practices for Design

Rules and Practices for Development

Rules and Practices for Testing

Planning

The XP methodology proposes planning as a dialogue between the parties involved in the project, including the client, programmers and coordinators or managers. The project begins compiling "User stories", which replace the traditional "use cases". Once the "User stories," programmers evaluate quickly the development time of each one.

If any of them has "risks" that do not allow to establish the complexity of development with certainty, small test programs ("spikes") to reduce these risks. Once these estimates, a planning meeting is organized, with the various stakeholders of the project (client, developers, managers) for the purpose of establishing a delivery plan or schedule ("Release Plan") in which everyone agrees. Once agreed this schedule, an iteration phase begins, where in each one of them is developed, tested and installed few "user stories".

According to Martin Fowler, plans in XP differ from traditional methodologies in three aspects: Simplicity of the plan. It is not expected that a plan requires a “guru” with complicated systems of project management.

The plans are made by the same people that will carry out the work. Plans are not predictions of the future, but simply the best estimate of how the plans are useful, but need to be changed when circumstances so require. From another way, it ends in situations in which the plan and reality do not match, and in these cases, the plan is totally useless.

The basic concepts of this planning are the following:

- User Stories

User stories are the technique used in XP to specify the requirements of the software. These are paper cards in which the client briefly describes the characteristics that the system must possess, be functional requirements or non-functional.

The treatment of user stories is very dynamic and flexible, at any time user stories can be broken, replaced for more specific or general ones, to be added new or modified. Every user story is sufficiently comprehensible and delimited so that programmers can implement it in a few weeks.

Regarding the information contained in the user history, there are several templates but there is no consensus on this. In many cases it is only proposed to use a name and a description or just a description, more perhaps an estimate of effort in days. Beck in his book presents an example of a customer story and task card in which the following contents:

Date,

Type of activity (new, correction, improvement),

Functional test, history number,

Technical and customer priority,

Reference to another previous history,

Risk,

Technical estimation,

Description,

Notes and a follow-up list with the date,

State things to finish and comments.

Release Plan

The delivery schedule states that user stories will be grouped to conform a delivery, and the order of the same. East schedule will be the result of a meeting between all project stakeholders (client, developers, managers, etc.). XP calls this meeting "Game of Planning", but can be denominated in a way that is most appropriate type of business and customer (eg Meeting of planning, planning meeting or planning workshop ") Typically the client will order and groups according to their priorities the histories of user. The delivery schedule is made in basis of the development time estimates made by the developers.

After some iterations it is recommended to hold a meeting with the actors of the project, to re-evaluate the plan of deliver and adjust if necessary.

Iteration Plan

User stories selected for each delivery is developed and tested in a cycle of iteration, according to the pre-established order. At the beginning of each cycle, a meeting is held of iteration planning.

Each user story is translated into programming tasks. Also, for each user history, establish the acceptance tests. These tests are performed at the end of the cycle in that develop, but also at the end of each one of the following cycles, to verify that subsequent iterations have not affected the above. Acceptance tests that have been failed in the previous cycle are analyzed for evaluating its correctness, as well as to predict that it will not happen again.

Daily follow-up meetings

The goal of having daily meetings is maintaining communication between the equipment, and share problems and solutions. In most these meetings, most of the participants simply listen, without having much to contribute. In order not to take unnecessary time out of the equipment, researchers suggests holding these meetings in a circle and standing.

Design

The XP methodology places particular emphasis on simple and clear designs. The most important in this methodology are the following:

Simplicity:

A simple design is implemented more quickly make it a resort. That's why XP proposes implement the simplest possible design that work. It is suggested never to implement functionalities that correspond to the iteration in which it is working.

Solutions "spike"

When technical problems arise, or when it is difficult to estimate the time for implement a user history, they can small test programs (called Spike), to explore different solutions. These programs are for testing or evaluation only. Solutions are often discarded after their evaluation.

Recoding

Refactoring consists of write again part of the code of a program, without changing its functionality, to make it simpler, more concise and / or understandable Many times, after finishing writing a program code, we think that if we started again, we would have done it in different way, more clearly and efficiently. However, as it is already ready and "works", it rarely is rewritten.

XP methodologies suggest recoding whenever necessary. While it may seem an unnecessary waste of time within the immediate, the results of this practice have fruits in the following iterations, whenever necessary to expand or change functionality. The philosophy that is being pursued is, as already mentioned, try to keep the code as simple as possible implement the desired functionality.

Metaphors

A "metaphor" is something that everyone understands, without the need for further explanation. The XP methodology suggests using this concept as an easy way to explain the purpose of project, and guide the structure and architecture of the same. For example, it can be a guide for the nomenclature of methods and classes used in the design of the code. Have clear names, which do not require further explanation, it a time saver.

It is very important that the client and the group of developers agree and share this "metaphor", so that they can dialogue in a "same language". A good metaphor should be easy to understand for the customer and in turn must have sufficient content to serve as a guide to the project architecture. However, this practice is often difficult to achieve.

CONCLUSIONS AND FUTURE WORK

It is not possible to expect to have a single development model suitable for all contexts of development. For example, some of the process limitations have been studied. Agile and, of course, it is understood that these are not the solution for all contexts. For this before proposing the specific development practices that team, its context must be defined and the best practices must be sought.

All proposals for development processes must have in addition to the definition and documentation of the processes that compose it (what), the set of methodologies (how) must be carried out. Within the proposal we present a set of defined development processes and documented, accompanied by a set of principles that must be with agile projects to adapt its development model.

The adaptation of development processes is indispensable to achieve continuous improvement and to fit the development context. Similarly, the use of support tools processes is essential in the inclusion of development equipment.

Some researchers believe that this proposal should be accompanied by a process to help organizations introduce process practices into their development. At this moment researchers are validating in several projects an implementation plan and researchers expect to have concrete results in the medium term.

CHAPTER 8

CONCLUSIONS, ENHANCEMENTS AND EXTENSIONS

Ivan Stanimirović

This chapter presents the conclusions of both the project and the book, extracted after the work done. And finally they are described improvements and extensions that would lead to the continuity of draft.

From the technological point of view, it has been seen that the implementation of an agile methodology provides increased equipment performance. On the other hand the implementation of the development pipeline has allowed the development team members to focus on more interesting tasks. The inclusion of operations and systems in Scrum methodology is to establish a precedent so that implementation become a referent.

It has also contributed to the community by maintaining infrastructure through the correction of an error, specifically of the Facter module of

Puppet. The developed code is in the stable version of the software as it has been explained in Chapter 3.

It has been proven that it is possible of managing a cloud based on Windows using Open Source systems running under Linux. In addition to systems with licensing of a different nature, management can be carried out through a single system.

Lastly, all the objectives set out in this book, not just providing a powerful cloud management tool, but also designing and implementing a development pipeline adapted to the continuous model delivery giving the ability to make frequent updates to the service.

The following describes the proposals for improvements that will give continuity to the project.

Robustness of the pipeline

Once a first version of the development pipeline is implemented, it should be carried out an analysis of its robustness in order to reinforce its weak points. The critical pipeline development systems should be redundant, both the server Jenkins as the Puppet Master, since before a fall of these the flow of the pipeline would be interrupted.

Non-disruptive updates

In collaboration with the platform development team, a mechanism by which the nodes can be warned to go to maintenance mode. In this way, nodes would not accept more incoming users and when current users have just run would proceed to the update.

Versioning of the infrastructure

Chapter 6 has explained how the software is versioned within the development based on the continuous delivery model. The configurations described in this book are versioned in a Git repository. In the same way, version of the infrastructure architecture, and even of correspondences between versions of the platform and the infrastructure and settings.

Orchestration of other clouds

This book is focused on the application cloud, but there are also other around the cloud, such as the cloud database, business intelligence, or the cloud metrics. All these clouds also require some management and maintenance, concept called cloud orchestration. Therefore, they could be included in orchestration to end up treating all the clouds as if it were just one.

Deployments through user interface

The purpose of this extension is to facilitate maintenance, making it easier and to accelerate the incorporation of new equipment components, since they would be less familiar with deployments. This extension would develop a control panel where the different production environments and the version of the platform that they were running, so that with a simple click could launch the entire deployment mechanism to a specific version for each environment.

CHAPTER 9

HORA: ARCHITECTURE-AWARE ONLINE FAILURE PREDICTION

Teerat Pitakrat [a], Dušan Okanović [a], André van Hoorn[a], Lars Grunske [b]

[a] University of Stuttgart, Institute of Software Technology, Reliable Software Systems, Germany [b] Humboldt University Berlin, Department of Computer Science, Software Engineering, Germany

ABSTRACT

Complex software systems experience failures at runtime even though a lot of effort is put into the development and operation. Reactive approaches detect these failures after they have occurred and already caused serious consequences. In order to execute proactive actions, the goal of online failure prediction is to detect these failures in advance by monitoring the quality of service or the system events. Current failure prediction approaches look at the system or individual components as a monolith without considering the architecture of the system. They disregard the fact that the failure in one component can propagate through the system and cause problems in other components. In this paper, we propose a hierarchical online failure prediction approach, called HORA, which combines component failure predictors with architectural knowledge. The failure propagation is modeled using Bayesian networks which incorporate both prediction results and component dependencies extracted from the architectural models. Our approach is evaluated using Netflix's server-side distributed RSS reader application to predict failures caused by three representative types of faults: memory leak, system overload, and sudden node crash. We compare HORA to a monolithic approach and the results show that our approach can improve the area under the ROC curve by 9.9%.

Keywords: Online failure prediction, Reliability, Component-based software systems

INTRODUCTION

Quality of service (QoS) problems in software systems at runtime, such as performance degradation and service outage, can lead to frustrated customers and losses in revenue. Software service providers need to make sure that the system satisfies its QoS requirements, e.g., in terms of response times not exceeding a certain threshold. During operation, such QoS properties are continuously monitored at the system boundary to assess QoS compliance. Based on Avižienis et al. (2004), we use the term failure in this paper to refer to the state of the component or system violating expected QoS levels.

Online failure prediction (Salfner et al., 2010) aims to foresee looming QoS problems at runtime before they manifest themselves. Accurate failure predictions are a prerequisite for preemptive maintenance actions, reducing the effect of problems or even completely preventing them from occurring (Brun et al., 2015; Calinescu et al., 2011; Candea et al., 2004; Li and Lan, 2006).

Existing online failure prediction approaches predict failures either of the whole system or of specific parts of the system. These approaches employ monolithic models which view the whole system or the service as one entity and predict failure events based on externally observable measures, e.g., response time (Amin et al., 2012a; Cavallo et al., 2010), event logs (Pitakrat et al., 2014a; Salfner and Malek, 2007), or system metrics (Baldoni et al., 2012; Williams et al., 2007). However, without considering the architecture or dependencies between components, these approaches are able to predict only the component failures but not their consequences on other parts of the system.

When faced with complex software systems comprised of a large number of internal and external components, the existing approaches may not be able to properly analyze all the influencing measures. For instance, system failures, which are visible to users, usually originate from complex interactions of erroneous components inside the system. These internal errors, which can be regarded as failures on component level, can propagate to other parts of the system through the architectural dependencies. This causes a chain of errors up to the system boundary and results in a failure on the system level (Avižienis et al., 2004; Cortellessa and Grassi, 2007; Nygard, 2007; Hiller et al., 2001; Johansson and Suri, 2005). This concept of failure propagation follows the definition by Avižienis et al. (2004). The notion of fault, error, and failure depends on the perspective of the viewer. For example, a failure from a component perspective can be considered as an error from the system perspective.

To overcome the challenges of predicting failures in complex systems, we hypothesize that online failure prediction can be improved by including architectural information about software systems. In this paper, we propose a hierarchical online failure prediction approach, called Hora.[1] The core idea is to combine architectural models with component failure prediction techniques. We use two types of architectural models: 1) Architectural Dependency Model (ADM) and 2) Failure Propagation Model (FPM). The ADM captures the dependencies between architectural entities. The FPM employs Bayesian network theory (Bishop, 2006) to represent the propagation paths of failures with corresponding probabilities. Both ADM and FPM can be automatically generated from an existing architectural model. At runtime, the FPM is constantly updated with individual failure probabilities, which are computed by component failure predictors. These predictors apply techniques, such as time series forecasting or machine learning, on monitoring data obtained from each component to predict component failures. In the last step, the FPM is solved to combine individual and propagated failure probabilities. The concept of separating the prediction into component failure prediction and the propagation prediction allows suitable prediction techniques to be used (and reused) for different types of components.

Our evaluation investigates the prediction quality of HORA with respect to the following research question: RQ1: Does HORA improve the prediction quality compared to a monolithic approach? If yes, to what degree? RQ2: How do the parameters of HORAaffect the prediction quality? RQ3: How does ADM affect the prediction quality? RQ4:What is the runtime overhead of HORA? The evaluation includes three typical types of faults, which are memory leak, system overload, and node crash, applied to Netflix's distributed RSS reader application. The results of our approach are compared to those of a monolithic approach, which does not consider architectural knowledge for the prediction but rather only the failure probability of each individual component. The results show that our approach can predict runtime failures with a higher prediction quality, with respect to standard evaluation metrics (Salfner et al., 2010).

To summarize, the paper contributes a novel online failure prediction approach, called HORA, which employs architectural knowledge together with a configurable and reusable set of failure prediction techniques. The approach is accompanied by a proof-of-concept implementation. The evaluation shows HORA's benefits over a monolithic approach.

This paper is an extended version of our previous work (Pitakrat et al., 2016). The approach is extended by an automatic extraction of the ADM. Moreover, this paper presents a comprehensive evaluation of the approach with respect to the previously mentioned research questions.

The remainder of the paper is structured as follows. Section 2 emphasizes the challenge of online failure prediction in distributed software systems. Section 3 details the HORA approach. The evaluation and the discussion of the results are presented in Section 4. Section 5 describes the related work. Finally, Section 6 draws the conclusion and outlines future work. Additionally, we provide supplementary material for this paper.

MOTIVATING EXAMPLE

Fig. 1 provides a high-level view on a typical distributed enterprise application system (Fowler, 2002). The example system conforms to the common three-tier architectural style of enterprise applications. Each of the tiers comprises a number of instances, to which requests are distributed over load balancers. Each instance comprises a complex stack of software architecture, middleware services, operating system, virtualization, and hardware components.

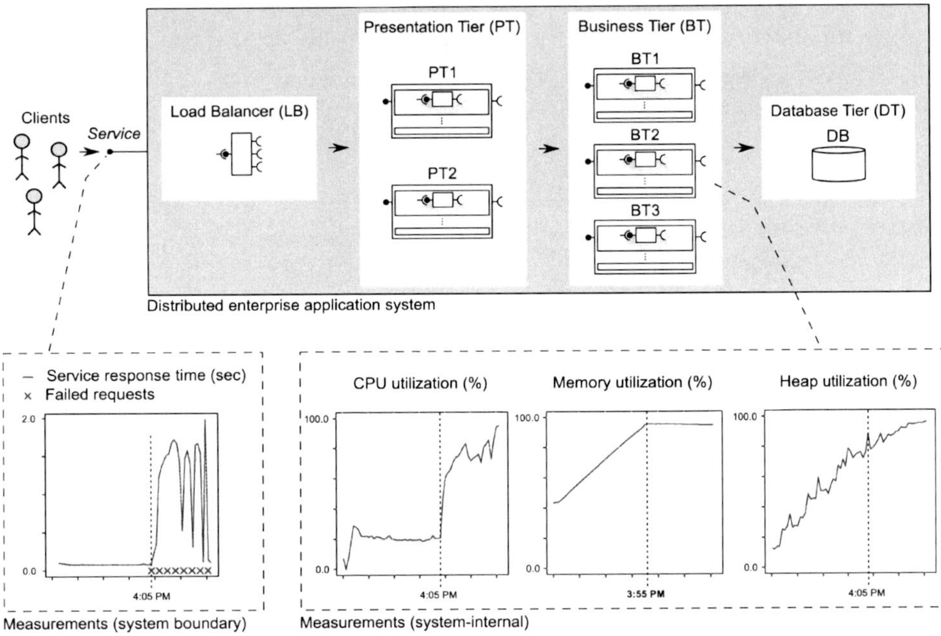

Figure 1: Running example: high-level three-tier architecture and selected measurements.

In this example, it can be observed that at 4:05 PM QoS problems manifest themselves at the system boundary as a prompt increase in response times and failing requests for the provided service. Online failure prediction approaches aim to predict failures before they occur in order to allow timely actions, such as preventive maintenance, to decrease or completely prevent system downtime. However, in this case, neither of the two metrics measured at the system boundary gives an indication about the upcoming problem. The traditional approaches for online failure prediction, e.g., time series forecasting based on service response time, are not appropriate in this case because the data does not contain any symptom that precedes the failure.

In addition to the system architecture, Fig. 1 includes three system-internal measures of the business-tier instance BT2, namely the utilization of CPU, system memory, and heap space of the Java Virtual Machine (JVM). It can be observed that the CPU utilization increases abruptly at 4:05 PM—the same time as the increase of the service response time. The utilization of system memory increases linearly until 3:55 PM when it reaches a level close to 100% and remains stable. The JVM heap space utilization shows an increasing trend until reaching almost 100%. In this scenario, we can conclude that the increase of the response times is caused by the increase of the CPU utilization. The increase of the CPU utilization is in turn caused by garbage collection activity inside the JVM—a common problem in Java systems. In this scenario, the root cause of the failure could be a memory leak in the BT2, which causes a chain of errors (Nygard, 2007) that propagates to the end users.

The online failure prediction approach, introduced in this paper, aims to predict this kind of problems by incorporating the failure probabilities of the internal components (in this case CPU, memory, and JVM) along with the failure propagation paths through the software system architecture.

THE HORA PREDICTION APPROACH

The main idea of HORA is that if the failure of each component in the software system can be predicted and the dependencies among the components are known, the consequence, i.e., the propagation, of the failures can also be predicted.

The HORA approach is comprised of two integral parts: two types of architectural models (Section 3.1) and hierarchical online failure predictions based on these models (Section 3.2).

Architectural models and transformation

This section introduces two types of architectural models used in HORA, namely Architectural Dependency Model and Failure Propagation Model, as well as the ADM extraction and transformation.

Architectural Dependency Model

There are already a number of existing architectural modeling languages and model extraction mechanisms (Lorenzoli et al., 2008; Schmerl et al., 2006), e.g., PCM (Becker et al., 2009), Descartes (Brosig et al., 2014), and SLAstic (van Hoorn, 2014).

However, these models are designed for different purposes, i.e., performance modeling, resource allocation, and capacity management. The information regarding component dependencies is not directly available and still needs to be extracted from these models. Thus, we introduce Architectural Dependency Model (ADM)—an intermediate model representing only the dependencies between architectural entities as needed by HORA.

The Architectural Dependency Model (ADM) relates every component to every other component along with the corresponding degrees of dependencies. It can be viewed as a $n \times n$ matrix for a system with n components. Each element in the matrix represents how much one component requires another component to function correctly. In other words, it is the probability that a failure of one component will affect another one based on the QoS metrics. For example, an element a_{ij} in the matrix represents the degree to which component c_i depends on c_j, where $1 \leq i, j \leq n$ and $i \neq j$. The ADM acts as an intermediate step which contains the information regarding the failure dependency and allows us to focus on one type of model.

Table 1 shows the table representation of the ADM for the example introduced in Section 2. Following a topological order, the database (DB) has no dependencies to other components; the business-tier instances BT1–3 depend on DB; the presentation-tier instances PT1–2 depend on the business-tier instances BT1–3; and the load balancer depends on PT1–2. Additionally, Table 1 includes the weights associated with these dependencies—in this case, assuming that requests among the tiers are equally load-balanced to the instances of the next tier. Note that for the sake of simplicity, we consider the six nodes from the example as monolithic components. For realistic scenarios (e.g., in the evaluation in Section 4), these components can be further decomposed into software and hardware components with measures, such as service response time, method response time, or resource utilization. Moreover, the network devices, e.g., routers that connect these components, can also be included to model connector failures.

Table 1: Table representation of the ADM for the system in Fig. 1.

Component	**Required components and weights**
DB	{}
BT1	{(DB, 1.0)}
BT2	{(DB, 1.0)}
BT3	{(DB, 1.0)}
PT1	{(BT1, $0.3\overline{3}$), (BT2, $0.3\overline{3}$), (BT3, $0.3\overline{3}$)}
PT2	{(BT1, $0.3\overline{3}$), (BT2, $0.3\overline{3}$), (BT3, $0.3\overline{3}$)}
LB	{(PT1, 0.5), (PT2, 0.5)}

Automatic extraction of Architectural Dependency Model

The ADM can be manually created by experts who have knowledge about the dependencies of the system. However, in practice, most systems contain a large number of components. Thus, the manual creation of the model becomes tedious, time-consuming, and error-prone.

In this section, we introduce an automatic extraction of the ADM from an existing architectural model. The architectural model used as the source is the SLAstic model (van Hoorn, 2014) which contains information regarding the structure and behavior of the system, and can be automatically extracted from monitoring data. The monitoring data required for the automatic model extraction comprises system-level resource usage, e.g., CPU utilization and memory consumption, and detailed execution traces of the application for each user request. From this data, the hardware and software components are discovered and linked together according to their observed relationship. The result of this automatic extraction is a SLAstic model which includes software components and their relationships, deployment information, and number of invocations of each component. This model extraction is described in detail in our previous work (van Hoorn, 2014). By combining the information from the SLAstic model, we obtain sufficient knowledge about the dependencies of the components to create the ADM. Fig. 2 visualizes the number of invocations between components extracted from the SLAstic model based on the running example in Fig. 1. The degree of dependency or dependency weight, denoted as w, between two software components A and B can be computed as

$$w_{AB} = \frac{i_{AB}}{i_A} \quad (1)$$

where i_{AB} is the number of invocations of component B by component A and i_A is the number of invocations of component A. In Fig. 2, PT1 is invoked 300 times and BT2 is invoked by PT1 100 times. Thus, the degree of dependency between PT1 and BT2 is 100300=0.33¯. Since the calling probability of component A to component B is 0.33¯,we hypothesize that if BT2 fails, the probability of PT1 failing due to the cascading failure is also 0.33¯. The degree of dependencies of other components can be computed in the same manner and the resulting ADM is similar to that in Table 1.

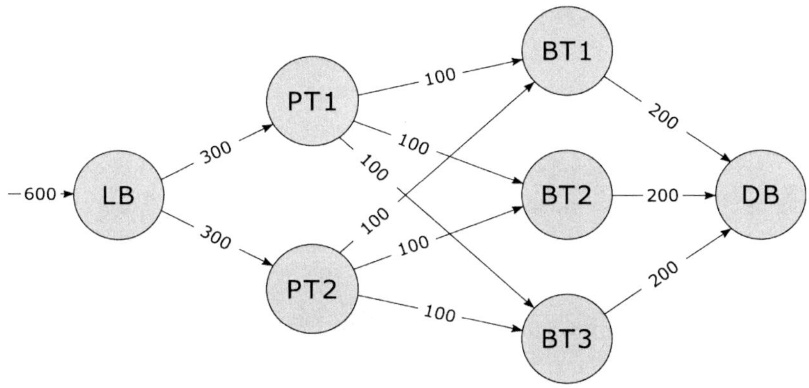

Figure 2: Calling relationship and number of invocations (denoted by arrows and the corresponding numbers) extracted from the SLAstic model.

In addition to the dependencies between software components, there are also dependencies between software and hardware components. For example, software components that are deployed on a physical node require that the hardware components of that node operate correctly. Each hardware component includes a set of measures of that component, such as utilization or temperature. In our experiment (Section 4), we consider the dependencies to the load average, CPU utilization, memory utilization, and swap utilization of all physical machines. For simplicity reasons, the hardware components are not shown in the example in Table 1. The automatic extraction of the ADM can be configured so that some software or hardware components can be excluded. Moreover, the degrees of dependencies between components can also be adjusted. This is useful in cases when a component appears in the model due to the monitoring setup but it is known not to have any effect on the QoS and the failure propagation.

Failure Propagation Model

Although the ADM contains the component dependencies and the corresponding weights, it is not suitable for failure prediction because the probabilities of cascading failures are not explicitly included. To model these probabilities, we transform the ADM into another representation, called Failure Propagation Model (FPM), which is an abstraction that represents the concept of the inference of failure propagations.

The FPM employs the formalism of Bayesian networks (Bishop, 2006) which is a probabilistic directed acyclic graph that can represent random variables and their conditional dependencies. These conditional dependencies are essential for the inference of failure propagation. The inference that includes only the health of the immediate dependent components will not be able to predict failures that propagate through many layers of the system. For instance, without conditional dependencies, the failure of DB in Fig. 1 would not have any effect on the failure probability of LB. Thus, conditional dependencies take into account the failure probabilities of all components in the dependency chain for the inference.

Fig. 3 depicts the Bayesian network which illustrates the relationship of node failures for the example system. Each node in the graph represents a software/hardware component and an arrow represents a causal relationship between components. The relationship implies that a failure of a parent component can cause a failure in the child component. For simplicity reasons, in this example we only consider each physical machine as a node in the graph without going into the details of each machine.

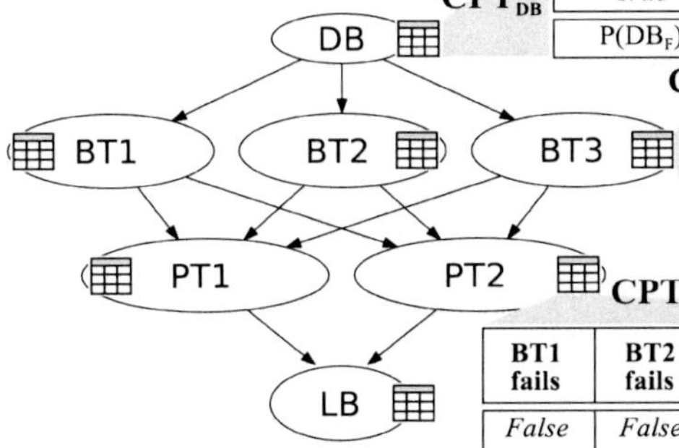

CPT$_{DB}$

DB fails	
True	*False*
P(DB$_F$)	1 - P(DB$_F$)

CPT$_{BT3}$

DB fails	BT3 fails	
	True	*False*
False	P(BT3$_F$)	1 - P(BT3$_F$)
True	1.00	0.00

CPT$_{PT2}$

BT1 fails	BT2 fails	BT3 fails	PT2 fails	
			True	*False*
False	*False*	*False*	P(PT2$_F$)	1-P(PT2$_F$)
False	*False*	*True*	$0.3\overline{3}$	$0.6\overline{6}$
False	*True*	*False*	$0.3\overline{3}$	$0.6\overline{6}$
False	*True*	*True*	$0.6\overline{6}$	$0.3\overline{3}$
True	*False*	*False*	$0.3\overline{3}$	$0.6\overline{6}$
True	*False*	*True*	$0.6\overline{6}$	$0.3\overline{3}$
True	*True*	*False*	$0.6\overline{6}$	$0.3\overline{3}$
True	*True*	*True*	1.00	0.00

Figure 3: Failure Propagation Model for the system in Fig. 1 with selected CPTs.

The conditional dependencies between the nodes in the graph are represented by a Conditional Probability Table (CPT). Each node in the graph has a corresponding CPT which contains conditional probabilities of possible failures occurring, given the failure probability of its parent components. For instance, the database is a node that does not depend on any other nodes. Therefore, its CPT contains only two failure probabilities that represent the probability of failure occurring, and not occurring, from inside the database itself. The table is shown in Fig. 3 as CPT$_{DB}$. The failure probability is denoted by P(DB$_F$) which is computed regularly at runtime by the corresponding component failure predictor, as will be detailed in Section 3.2.1. On the other hand, the operation of a business tier (BT1–3) requires a database (DB) with a dependency weight of 1.0, according to the ADM in Table 1. This means if the database fails, the business-tier instances will also fail. The CPT of the business-tier instance BT3 is presented in Fig. 3 (CPT$_{BT3}$). The first row indicates the failure probability of the business tier itself, if the database is operating properly. The second row indicates the probability of BT3 failing if the database fails.

A more complex relationship can be seen from the presentation tier which requires at least one business-tier instance. If one business-tier instance fails, the presentation tier can still operate by forwarding requests to the remaining business-tier instances. As listed in Table 1, the dependency weight of each presentation-tier instance to each business-tier instance is approximately 0.33. This implies that, for each business tier failure, the failure probability of the presentation-tier instances will increase by approximately 0.33. Hence, if all business-tier instances fail, this failure probability will sum up to 1.0 which means that the presentation-tier instances will also fail. The CPT of PT2 is presented in Fig. 3 as CPT_{PT2}.

Assuming that a component depends on n other components with $n \geq 1$, the CPT of a component c_0 can be expressed by a multiplication of a truth table matrix **T** of size $2^n \times n$ and the weight matrix Wc0 of size $n \times 2$:

$$\mathrm{CPT}_{c_0} = \mathbf{T} \cdot \mathbf{W}_{c_0} \tag{2}$$

where

$$\mathbf{T} = \begin{pmatrix} c_1 & \dots & c_{n-2} & c_{n-1} & c_n \\ 0 & \dots & 0 & 0 & 0 \\ 0 & \dots & 0 & 0 & 1 \\ 0 & \dots & 0 & 1 & 0 \\ 0 & \dots & 0 & 1 & 1 \\ 0 & \dots & 1 & 0 & 0 \\ \vdots & \ddots & \vdots & \vdots & \vdots \\ 1 & \dots & 1 & 1 & 1 \end{pmatrix}, \tag{3}$$

$$\mathbf{W}_{c_0} = \begin{pmatrix} w_{c_0 c_1} & 1 - w_{c_0 c_1} \\ w_{c_0 c_2} & 1 - w_{c_0 c_2} \\ \vdots & \vdots \\ w_{c_0 c_n} & 1 - w_{c_0 c_n} \end{pmatrix}, \tag{4}$$

with c_i, $1 \leq i \leq n$, are required components and wc0ci are the corresponding dependency weights from component c_0 to component c_i.

The CPTs of other nodes are also created in the same manner as those for the database, business tier, and presentation tier. The complete model with all CPTs is used as a core model to infer about the failure probability of each component and failure propagation. The failure prediction and inference of the model at runtime will be discussed in detail in Section 3.2.

Hierarchical online failure prediction

The prediction of the failures and the inference of their propagation at runtime are comprised of two main steps. The first step is the prediction of individual component failure probabilities (Section 3.2.1). The second step is the inference of failure propagation based on the individual component failures and the FPM (Section 3.2.2).

Component failure prediction

The purpose of component failure predictors is to predict failures of each individual component. Each predictor monitors the runtime measurements of one component and makes a prediction whether the current state might lead to a component failure.

At runtime when a component failure predictor produces a new prediction, the result is used to update the FPM to keep the model up-to-date. Since the prediction result of the component failure predictor indicates the probability of a failure occurring from the component itself, this probability then replaces the first row of the CPT of the corresponding node in the model. For example, if the predictor of BT3 predicts that it is going to fail with a probability of 0.8, the probabilities in the first row of CPT_{BT3} in Fig. 3, in which DB failure is False, will be set to 0.8 and 0.2, accordingly. This process is periodically performed for all component failure predictors. Fig. 4a illustrates the concept of a component failure predictor based on the memory consumption of business tier BT2 in Fig. 1. Since the memory consumption is time series data, we can, for instance, employ autoregressive integrated moving average (ARIMA) (Shumway and Stoffer, 2013) as a component failure prediction technique. The goal of the prediction is to predict when the memory utilization will reach the 100% threshold, assuming that the machine will have a performance degradation when the memory is depleted, which can cause a service failure. The thin solid line in the graph indicates the monitoring data of memory utilization up to 3:35 PM. The dash-dotted line indicates the prediction of the memory utilization in the next 20 min with a prediction interval in light grey.

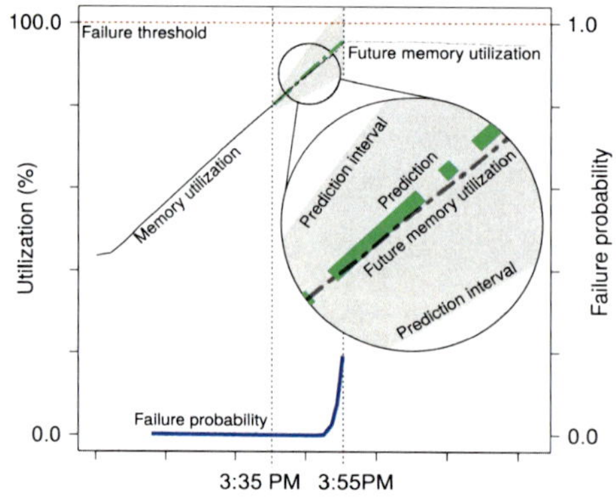

(a) Memory failure prediction (monolithic approach)

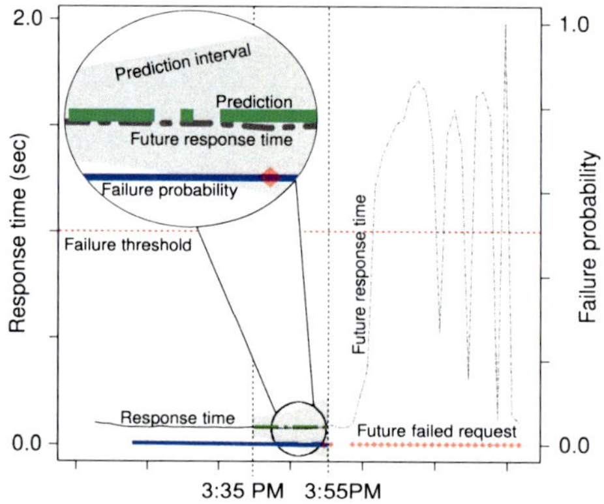

(b) Service response time failure prediction (monolithic approach)

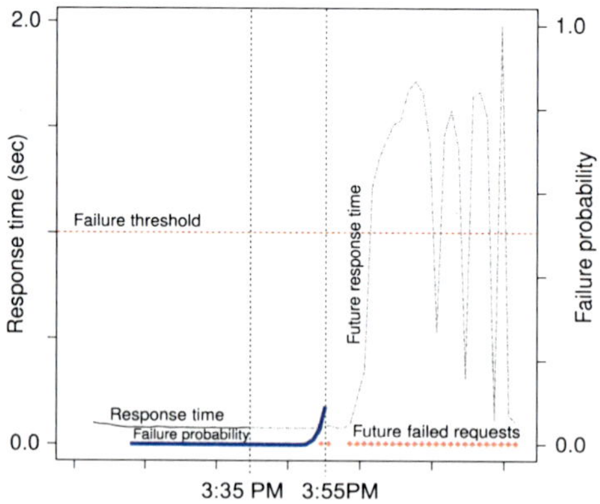

(c) Service response time failure prediction (HORA)

Figure 4: Example prediction results: monolithic approach vs. HORA.

The probability of the monitoring data crossing the failure threshold α can be computed using the probability density function f(x) of the predicted performance measure:

$$P(X > \alpha) = \int_{\alpha}^{\infty} f(x)dx \quad (5)$$

Fig. 5 depicts the probability density function of the memory utilization at 3:55 PM. Assuming that the input data is normally distributed, the prediction error is also normally distributed (Montgomery et al., 2009). Thus, the prediction interval assembles a normal distribution. The predicted value of 93% indicates the mean of the distribution. The 95% prediction interval covers the ±1.96σ area of the distribution (Montgomery et al., 2009). The probability of the memory utilization crossing the failure threshold at 100% can be computed using Eq. (5) with α=100. It is worth noting that, in the case of non-normally distributed data, the probability density function f(x) in Eq. (5) can be replaced by a suitable distribution to compute the component failure probability.

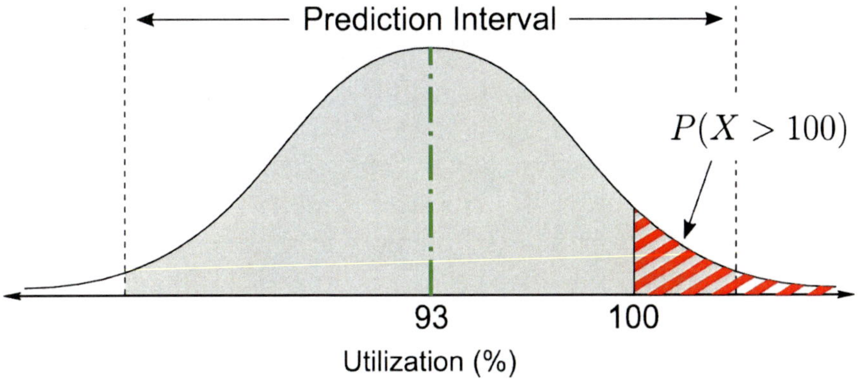

Figure 5: Probability density function of memory utilization at 3:55 PM and memory failure probability.

In this section, we show how HORA employs ARIMA as a component failure predictor. HORA is designed to utilize any other prediction method, such as other time series forecasting (Amin et al., 2012a) or machine learning techniques (Pitakrat et al., 2013; 2014b), as long as they can 1) predict the failure probability and 2) provide the expected time of the failure.

Inference of failure propagation

The inference of the failure propagation is the last step of HORA, which aims to predict what can be the effects of component failures. Once the component failure probabilities are updated in the model, we use Bayesian inference (Bishop, 2006) to obtain the failure probabilities of all components. The inference of the components' failure probabilities takes into account not only their own failure probabilities but also those of their parents and ancestors. If a node's ancestors have high failure probabilities and the corresponding degree of dependency w is also high, its failure probability will also be high. Therefore, the inference allows us to model and predict failure propagation from inside to the outside of the system. At runtime, the inference is done at regular intervals to provide the current failure probabilities of all components. The lead time of the inference is the same as those of the component failure predictors and can be set in the configuration of HORA.

Fig. 4b shows the result of an ARIMA component failure predictor which predicts service failures based only on the observed response time at the system boundary, i.e., the load balancer. The predictor does not take into account the failure probabilities of other components. It is obvious that this predictor cannot predict the first few occurrences of the service failure. This is because the response time starts increasing exactly when the first failure occurs. Fig. 4c illustrates the prediction of HORA which takes into account the architectural dependencies of the components. HORA considers not only the failure probability of the response time but also the failure probability of memory utilization in Fig. 4a and the Failure Propagation Model in Fig. 3. By comparing failure probabilities in Fig. 4a and c, it can be observed that the failure probability at 3:55 PM is scaled down from approximately 0.2–0.1. This effect is due to the weight of the dependency presented in Table 1. This inference provides a better estimation of the failure probability because the system contains three instances of the business tier and a failure in one instance does not necessarily have to cause a service failure. Should the prediction be made without the inference, the failure probability would be higher and cause a false alarm.

EVALUATION

This section describes the evaluation of HORA and aims to answer the following research questions:

RQ1: Does HORA improve prediction quality compared to a monolithic approach? If yes, to what degree?

This research question refers to our main hypothesis that we can improve the quality of online failure prediction using HORA. In order to investigate the degree of improvement, we compare HORA with a monolithic approach which is a set of component failure predictors that do not use the architectural knowledge to propagate the failure probabilities. The evaluation is conducted as a lab experiment with a distributed enterprise application under a synthetic workload with three different types of faults.

RQ2: How do the parameters of component failure predictors affect the prediction quality?

The configuration of HORA includes various parameters, such as aggregation window size, number of historical data, and lead time. When collecting response time of the service and components, the monitoring data is not time series because the functions are not executed at regular intervals. Thus, the data have to be preprocessed by computing the average response time in a time window, denoted here as aggregation window size. The historical data defines how much of the past data is used to predict future observations. The lead time is the time between when the prediction is made and when the failure is expected to occur. A longer lead time will allow the operator to be informed about the pending failures earlier than a short lead time.

This research question aims to investigate the sensitivity of HORA's prediction quality with respect to different parameter configurations.

RQ3: How does ADM affect the prediction quality?

A system can be represented by multiple ADMs. For example, different configurations of the monitoring result in different ADMs when using the automatic extraction described in Section 3.1.2. This research question aims to investigate the impact of the ADM with different numbers of components and degrees of dependencies on the prediction quality.

RQ4: What is the runtime overhead of HORA?

The goal of HORA is to predict failures at runtime. Therefore, the runtime overhead of HORA should be taken into account. This research question aims to investigate how much time is required for HORA to analyze the monitoring data and to infer the FPM to make predictions.

Experimental methodology and settings

Hora framework implementation

According to the concept presented in Sections 3.1 and 3.2, we have developed a Java-based proof-of-concept implementation. For the Bayesian network, we employ the Jayes library (Kutschke, 2013). The runtime measurements of the system, including application-level performance and execution traces, as well as resource utilization measurements, are collected by the Kieker monitoring framework (van Hoorn et al., 2012). For time series forecasting, the statistical library R (R Core Team, 2015) is used. HORA can be configured by using a configuration file. The configuration parameters relevant to the evaluation will be presented in Section 4.1.3. The implementation of HORA is available as part of the supplementary material.

System under analysis

The HORA approach is evaluated using an extended version of a distributed RSS feed reader application developed by Netflix.[2] This microservice-based application (Newman, 2015) provides a web service where users can view, add, or delete RSS feeds.

Our setup contains two instances in the presentation tier, three instances in the business tier, and one database. Additionally, the system has one frontend load balancer, one service discovery node, and two RSS feed servers. The workload driver is set up on a separate node and uses Apache JMeter (Halili, 2008) to generate user requests. The workload generated by each user includes view, add, and delete operations of the RSS feed with an average 3 sec think time between two requests. The number of concurrent users is set to 150 throughout the experiment. On average, the workload generates approximately 90 requests per second.

The described system is deployed on Emulab (Hibler et al., 2008), which is a large-scale virtualized network testbed. Each of the instances is a physical machine type pc3000 which is equipped with a 3 GHz 64-bit Xeon processor and 2GB of physical memory, running Ubuntu 14.04.1 LTS and Java 1.7.0 update 75.

Experiment configuration

This section describes the configurations of parameters and models that we use for the evaluation.

ADM and monitored variables

The ADM used in the evaluation is similar to that in Table 1. Other than the physical nodes in the system, we include additional software and hardware components with the following measures:

- Response times of view, add, and delete operations at the system boundary (frontend load balancer),
- Response times of methods involved in processing requests in all presentation- and business-tier instances,
- Load average, CPU utilization, memory utilization, and swap utilization of all physical machines.

Failure definition

A service failure is defined as an event that occurs when the service deviates from the correct service (Avižienis et al., 2004). For example, the deviation of the service can be regarded as an increase in the response time, a service outage, or an incorrect result. In our experiment, we classify a service to be in a healthy or failure state by observing the response time and the response itself in 2 min windows. We consider a service to fail if the following conditions occur within a certain time window: 1) the 95th percentile of the server-side response times of all of the requests in that window exceeds 1 s and/or 2) the ratio of successful requests over all requests falls below 99.99%.

The response time threshold of all methods is set to 1 s in the same manner as the server-side response time. We select this value because it eliminates the need of a training phase while still allowing the component failure predictors to make predictions. An alternative to this is to have the thresholds set manually. However, it is infeasible in practice when the system contains a large number of components. A second alternative is to determine the thresholds by learning the response times of all methods. However, this would introduce a learning phase and the response time could vary depending on the context in which the method would be used. In our future work, we plan to investigate the feasibility of prediction techniques that do not require the definition of a failure threshold, e.g., control charts.

The failure definition of other architectural entities are set according to the types of the entities. The memory utilization threshold is set to 100% according to its physical limit because after this point the operating system will start swapping which uses the space on the hard drive. The heap utilization threshold is set to 90%, since the garbage collector is triggered automatically when utilization becomes too high. The load average represents the number of tasks in the CPU queue over time and provides more information than the CPU utilization (Walker, 2006). For example, a 1 min load average of 1.0 means that there is one task in the CPU queue on average in the past minute. Thus, we set the failure threshold of the load average to 1.0.

Prediction technique

At runtime, the monitoring data, containing execution traces and resource measurements, are aggregated into windows of size 2 min, which are then pre-processed according to the type of architectural entity measure. The 95th percentile is calculated for the response time and method response time while the mean is calculated for the load average, memory utilization, and heap utilization.

As stated in Section 3.2, we use ARIMA as a component failure predictor. The size of the historical data for ARIMA is set to 10 min. The prediction lead time is 10 min with a 95% confidence level. The lead time of the FPM is the same as the component failure predictors, which is 10 min.

EVALUATION METRICS

This section introduces the metrics used to evaluate the quality of the prediction. A complete list of metrics and detailed descriptions can be found in Salfner et al. (2010).

Table 2 presents the four basic evaluation metrics as a contingency table. A true positive (TP) is a correct prediction of a failure. A false positive (FP) is a prediction of a failure that never occurs. A false negative (FN) is a miss which means the failure occurs but it was not predicted. A true negative (TN) is a correct prediction that a failure does not occur. Moreover, we consider four derived metrics, which are precision, recall or true-positive rate (TPR), false-positive rate (FPR), and accuracy, as listed in Table 3.

Table 2: Contingency table.

Prediction	**Actual**	
	Failure	**Non-failure**
Failure	True positive (TP)	False positive (FP)
Non-failure	False negative (FN)	True negative (TN)

Table 3: Selected derived evaluation metrics.

Metric	**Formula**
Precision	TPTP+FP
Recall, True-positive rate (TPR)	TPTP+FN
False-positive rate (FPR)	FPFP+TN
Accuracy	TP+TNTP+TN+FP+FN

A Receiver Operating Characteristic (ROC) curve (Fawcett, 2006) represents the quality of the prediction by relating (TPR) to FPRs for different prediction thresholds, as shown in Fig. 9. A perfect predictor has a curve that covers the entire area of the plot because the TPR is 1 while the FPR is 0. Thus, the closer the curve is to the (0, 1) point, the better the prediction is.

Area Under ROC Curve(AUC) measures the area that is covered by a ROC curve and allows comparison between different ROC curves. A perfect predictor would have an AUC of 1. The AUC is recommended to be used as a single-number metric for evaluating learning algorithms (Bradley, 1997; Huang and Ling, 2005). Thus, in our evaluation, the AUC is used as a representative metric for the comparison of the prediction quality of Hora and the monolithic approach. Furthermore, to evaluate the significance of the improvement, we use two-sided hypothesis testing

(Wohlin et al., 2012) to compare the AUC with the following null and alternative hypotheses:

H0:AUCHora=AUCMonolithicH1:AUCHora≠AUCMonolithic

The p-value is reported as the result of the test. The method used in testing is introduced by DeLong et al. (1988) to compare two or more AUCs. The ROC curves, AUCs, and the results of the tests presented in Section 4.2 are generated using the pROC package (Robin et al., 2011) available in R (R Core Team, 2015).

Fault injection

In our evaluation, we consider three types of faults from real world incidents (Pertet and Narasimhan, 2005), which are memory leak, system overload, and node crash. We inject one type of these faults into each experiment run and apply Hora and the monolithic prediction approach to predict the failures. Each experiment lasts two hours and is repeated 10 times. The reported evaluation metrics are obtained by combining and analyzing the raw prediction results of all runs. The details of each type of faults are described as follows.

Memory leak—In the experiment, we introduce a memory leak in one of the business tiers. Every time a request is sent from the presentation tier to this specific instance, 1024 bytes of memory will be allocated and never released.

System overload—System overloads occur when the workload increases, either gradually or abruptly, until the system is not able to handle all the incoming requests. In this scenario, instead of using a constant workload, we increase the number of users until service failures occur (detailed in Section 4.1.3).

Node crash—Unexpected node crashes are not uncommon in real systems. They can be caused by both software and hardware, such as operating system crashes, hardware failures, or power outages. We introduce this problem by intentionally shutting down two of the business tier instances at 90 and 95 min into the experiment.

Results

The following sections (Sections 4.2.1–4.2.4) present the results for the four research questions.

Investigation of prediction quality (RQ1)

In this section, we provide the results and explanation for the experiments with different types of faults. The detailed evaluation metrics of all scenarios are summarized in Table 4.

Table 4: Comparison of all evaluation metrics for the different types of faults.

Fault type	Prediction approach	Precision	Recall, TPR	FPR	Accuracy	**AUC**	**AUC improvement**	**p-value**
Memory leak	HORA	0.612	0.945	0.096	0.91	0.931	5.7%	$<2.22\times10^{-16}$
	Monolithic	0.84	0.758	0.024	0.945	0.881		
System overload	HORA	0.181	0.876	0.216	0.789	0.894	17%	$<2.22\times10^{-16}$
	Monolithic	0.352	0.564	0.059	0.92	0.764		
Node crash	HORA	0.059	0.73	0.298	0.703	0.771	< 0.1%	0.746
	Monolithic	0.209	0.582	0.085	0.902	0.766		
Overall	HORA	0.419	0.833	0.091	0.903	0.92	9.9%	$<2.22\times10^{-16}$
	Monolithic	0.475	0.692	0.065	0.916	0.837		

Memory leak—The memory leak in one of the business-tier instances causes the memory utilization to increase over time. As shown in Fig. 6a, at approximately the 52th min, the component failure predictor for memory in one of the business tier instances predicts that the memory utilization will cross the threshold at the 62th min with a high probability. This failure probability propagates to other parts of the business tier and, consequently, to the operations of the presentation tier and the load balancer, as can be seen in Fig. 6c and d. Since the memory leak is occurring in only one of the three business tier instances, the failure probability is reduced by the inference by a factor of 3. At approximately the 75th min, the garbage collector starts continuously freeing up heap space, which causes the load average to increase (Fig. 6b). The failure probability from the load average further increases the failure probability of the service and can be seen in Fig. 6c and d.

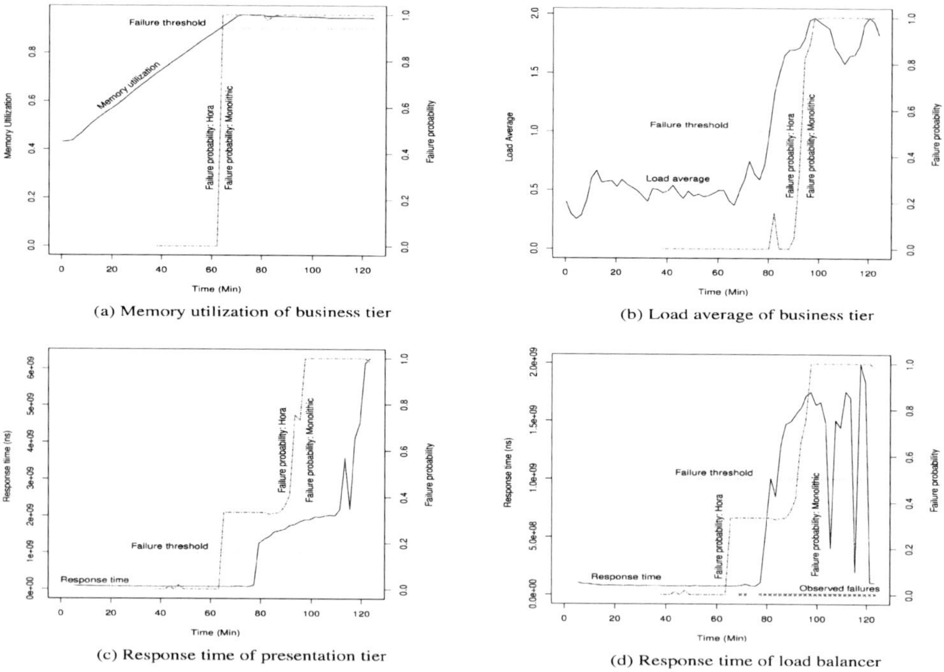

Figure 6: Timeline plots of selected components for memory leak scenario.

As can be expected, the memory leak causes a sudden increase in the service response time, which cannot be predicted by the response time predictor. On the other hand, HORA considers the memory utilization of the business tier and propagates the failure probability to the service boundary. The result shows that HORAcan predict the service failure 10 min before it occurs with a failure probability of approximately 0.3. The ROC curves of HORA and the monolithic approach are depicted in Fig. 9a.

System overload—The failures caused by overloading the system start occurring at approximately 80 min into the experiment. The increasing number of concurrent users causes the load average of the business-tier instances to exceed the failure threshold. As a result, some of the requests sent from the presentation tier to the business tier are rejected. After a pre-defined number of unsuccessful retries, the presentation tier responds with a page indicating that an error has occurred.

Fig. 7b depict the load average of one business tier instance which gradually increases over time, while the memory utilization remains stable in Fig. 7a. The failure probability from the load average propagates through the architecture to the presentation tier and load balancer, as seen in Fig. 7c and d.

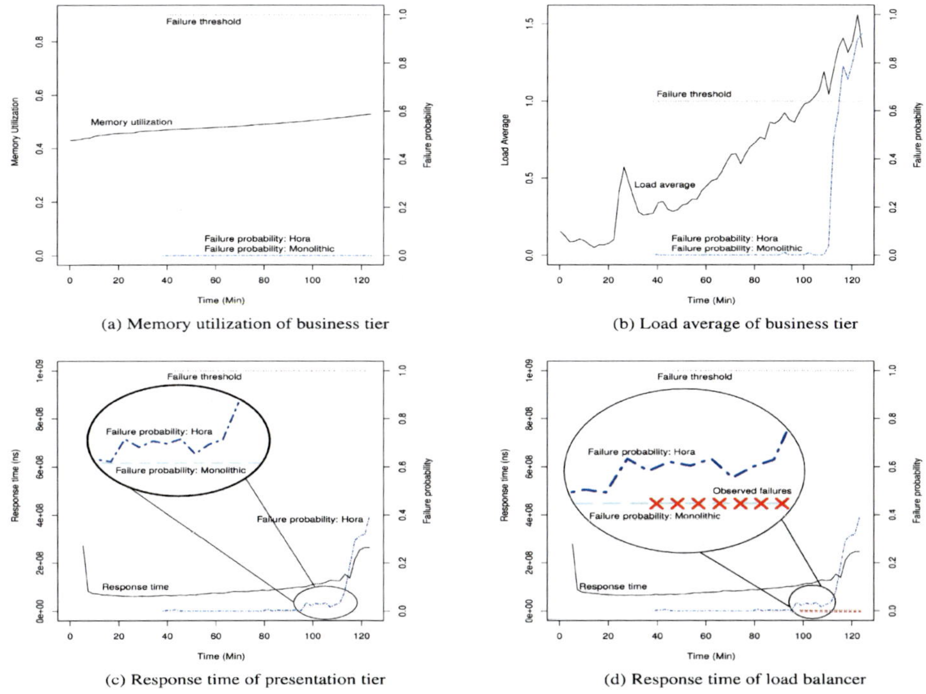

(a) Memory utilization of business tier

(b) Load average of business tier

(c) Response time of presentation tier

(d) Response time of load balancer

Figure 7: Timeline plots of selected components for system overload scenario.

The result shows that HORA can predict this type of service failure since it takes into account the dependency of the presentation tier on the business tier. On the other hand, the predictor that observes only the increase in the response time at the system boundary is not able to predict this type of service failure since the response time does not exceed the threshold. The ROC curves of both approaches are presented in Fig. 9b.

Node crash—In this scenario, we intentionally crash the second instance of the business tier at 90 min, and the third instance at 95 min into the experiment. The one remaining business tier instance has to take over the workload from those that failed. This causes the load average of the remaining instance to increase all of a sudden as can be seen in Fig. 8b. As a consequence, the service response time of the presentation tier and load balancer, shown in Fig. 8c and d, also increases unexpectedly.

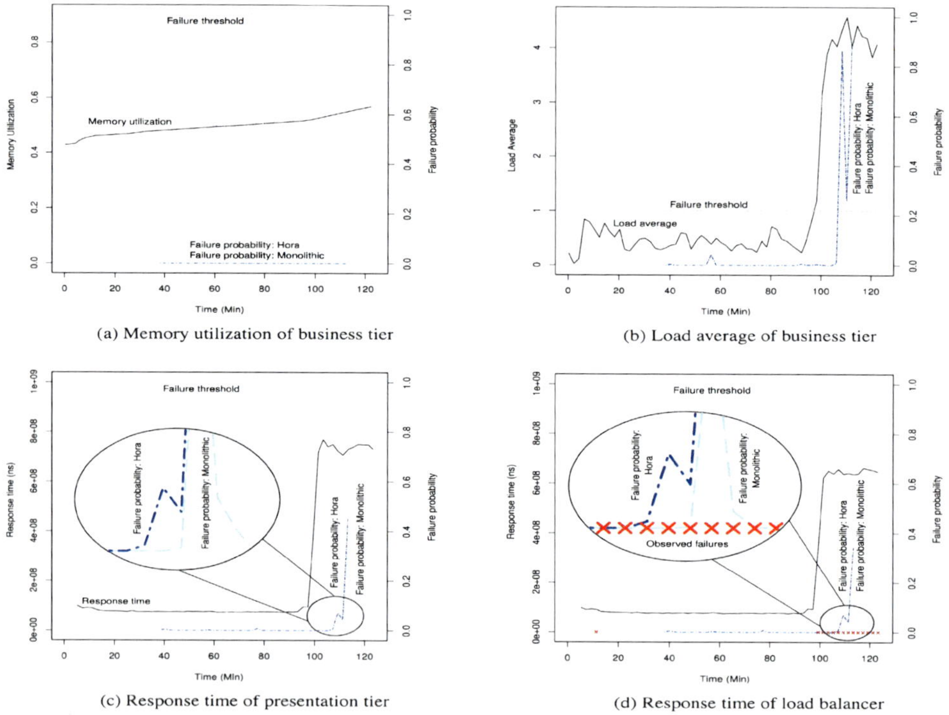

(a) Memory utilization of business tier

(b) Load average of business tier

(c) Response time of presentation tier

(d) Response time of load balancer

Figure 8: Timeline plots of selected components for node crash scenario.

The result in Fig. 9c shows that HORA performs as well as the monolithic approach. This is because the crash occurs unexpectedly without any preceding symptom. The component failure predictors cannot predict this and, therefore, the FPM does not have any failure information to propagate.

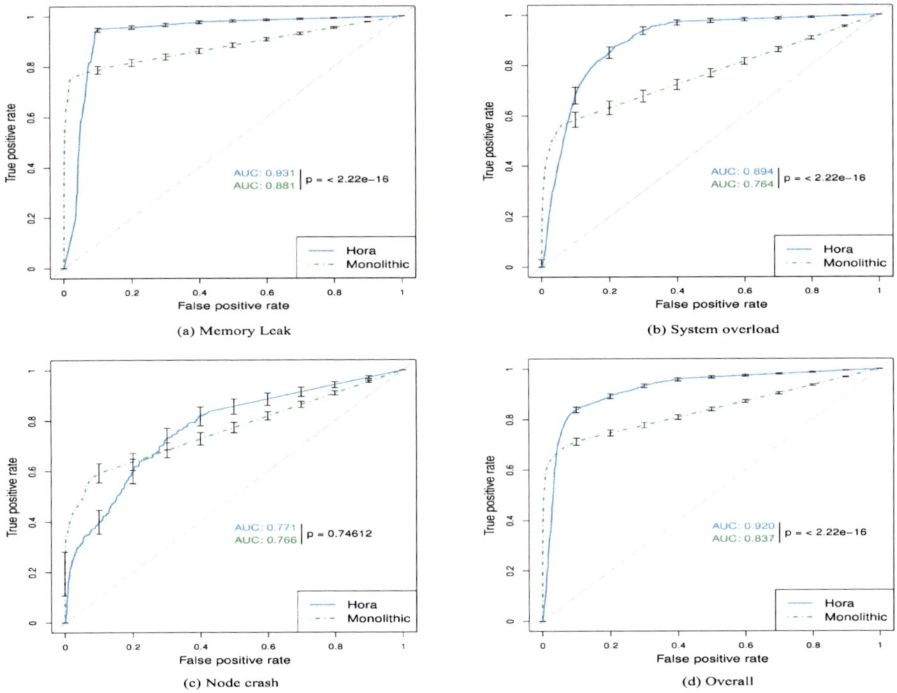

Figure 9: Comparison of ROC curves for the different types of faults.

Overall—We evaluate the overall prediction quality of Hora by analyzing the combined raw prediction data of all three scenarios. The results in Fig. 9d and Table 4 show that Hora improves the overall AUC by 9.9%, compared to the monolithic approach.

4.2.2. Investigation of parameter impact (RQ2)

We investigate three prediction parameters of the component failure predictors which are the aggregation window size, historical data size for ARIMA, and lead time. The reported metrics are computed by evaluating the raw prediction results of all fault types.

Aggregation window length—We vary the length of the aggregation window from 2 to 8 min. Fig. 10a shows that the precision increases when the window length increases from 2 to 6 min. However, when the length reaches 8 min, the precision, TPR, and AUC drop significantly. This is because small aggregation windows preserve the small variations in the data which causes the predictor to produce a lot of false positives. On the other hand, as the aggregation window gets larger, the variations in the data get lost. The predictor cannot detect the changes and, therefore, misses the failures.

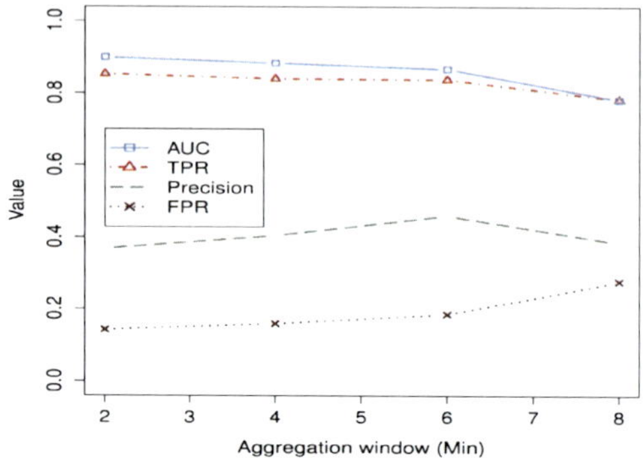

(a) Aggregation window length

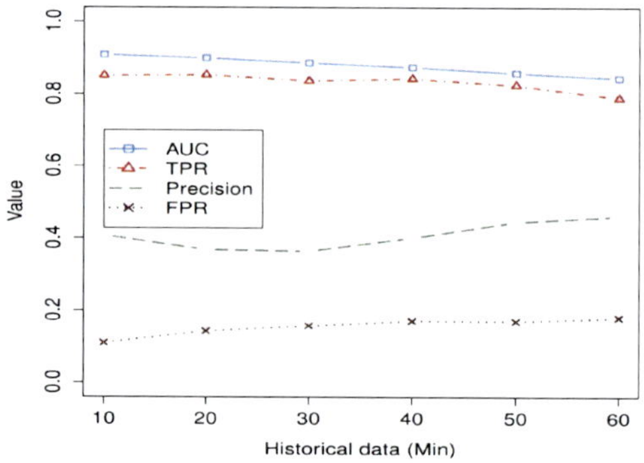

(b) Historical data length

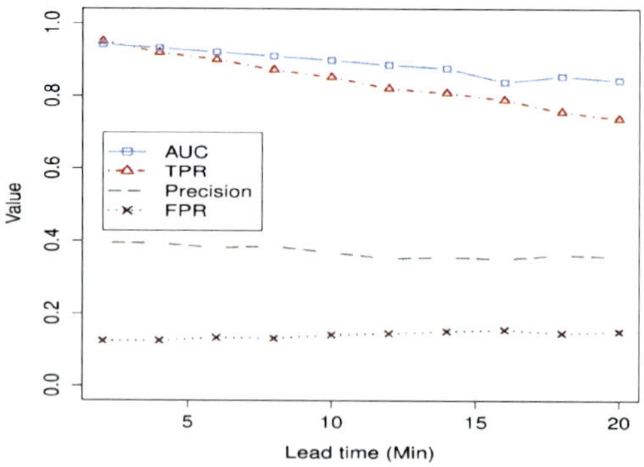

(c) Lead time

Figure 10: Prediction quality of HORA with different parameter configurations.

Historical data length—The length of historical data denotes how many data points further back in the past ARIMA considers for the prediction. Fig. 10b shows that the AUC and TPR exhibit a decreasing trend with the data length, while the FPR shows a slightly increasing trend. This is because the prediction quality is reduced when the predictors consider more data from the past because the short-term variations are neglected by the moving average part of ARIMA (Shumway and Stoffer, 2013).

Lead time—When increasing the lead time, the result in Fig. 10c shows a decreasing trend of precision, TPR, and AUC. The decrease is caused by the uncertainty of the prediction. As ARIMA is used to forecast the time series data, the confidence interval becomes larger as we make prediction further into the future. Therefore, choosing an optimal lead time depends on the application and the costs of a false positive and a false negative.

Investigation of Architectural Dependency Model impact (RQ3)

The automatic extraction of the ADM (Section 3.1.2) allows fine-tuning of the model, e.g., adjusting the degrees of dependencies and excluding some components. This section investigates the impact of the ADM configurations on the prediction quality.

Degree of dependency—The degree of dependency between software components can be directly computed from the number of invocations described in Section 3.1.2. However, the computation of the degree of dependency from software to some hardware measures are not straightforward. For example, the load average measure contains three different values, i.e., 1, 5, and 15 min averages. There are no clear guidelines what might be the effect on the operation of the physical machine if these measures exceed the threshold.

To investigate the impact of the degree of dependency, we vary the degree of dependency of the load average from 0.2 to 1.0. The results in Fig. 11a show that the AUC, TPR, precision, and FPR remain almost constant. This is because the failure probabilities are always propagated to other parts of the system by the FPM regardless of how small they are. Those small propagated probabilities at the system boundary still provide the signs that the failure is imminent which are sufficient to trigger the warning. Therefore, we can conclude that varying the degree of dependency does not significantly affect the prediction quality.

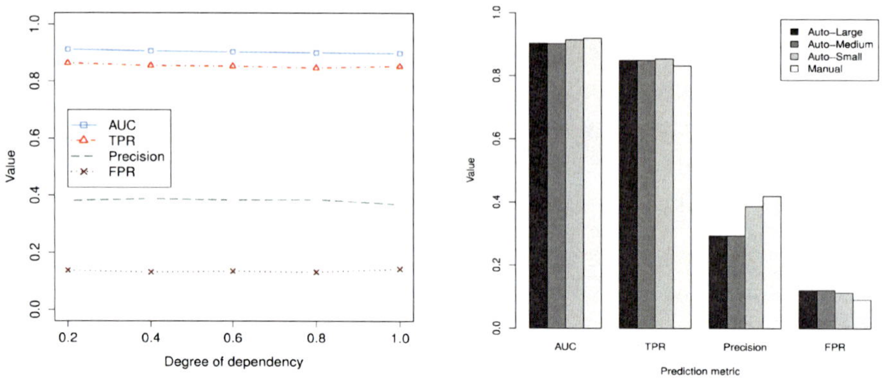

(a) Degree of dependency to load average

(b) Size of ADM

Figure 11: Prediction quality of HORA with different ADM configurations.

Size of ADM— We evaluate four different ADMs which contain different numbers of architectural components:

- Auto-Large—The model is automatically generated and contains 98 components which include software components, as well as CPU utilization, memory utilization, swap utilization, and load average of all physical nodes.
- Auto-Medium—The model is automatically generated and contains 80 components which include software components, as well as memory utilization, and load average. Compared to Auto-Large, the CPU utilization and swap utilization are removed.
- Auto-Small—The model is automatically generated and contains 56 components which include only important software components, as well as memory utilization, and load average. Compared to Auto-Medium, the intermediate software operations are removed.
- Manual—The model is manually created by system experts and contains 48 components which include only the important software components, as well as memory utilization, and load average.

The complete details of these four models can be found in the supplementary material. The evaluation results of the models are presented in

Fig. 11b. Although all ADMs have approximately the same AUC, TPR, and FPR, the manually created ADM achieves the highest precision. This is because the automatic extraction adds the components that do not improve the prediction and even decrease the precision. In other words, they do not help predict more failures, but rather produce more false alarms. However, the automatic extraction of the model has an advantage that it can create the ADM for a large system which is infeasible for a manual creation. Therefore, it is a trade-off between the ease of the model creation and the prediction quality. Nonetheless, the automatically generated model can still be fine-tuned by system experts to produce even better prediction results.

Investigation of runtime overhead (RQ4)

The prediction results are obtained by collecting the monitoring data from the system described in Section 4.1.2 and executing the offline analysis on a separate machine equipped with 3.10 GHz Intel Xeon E31220 running Ubuntu 12.04.5 LTS. Fig. 12illustrates the runtime overhead of HORA for the four different ADMs. It can be observed that the larger the model is, the more time it requires for the analysis and prediction.

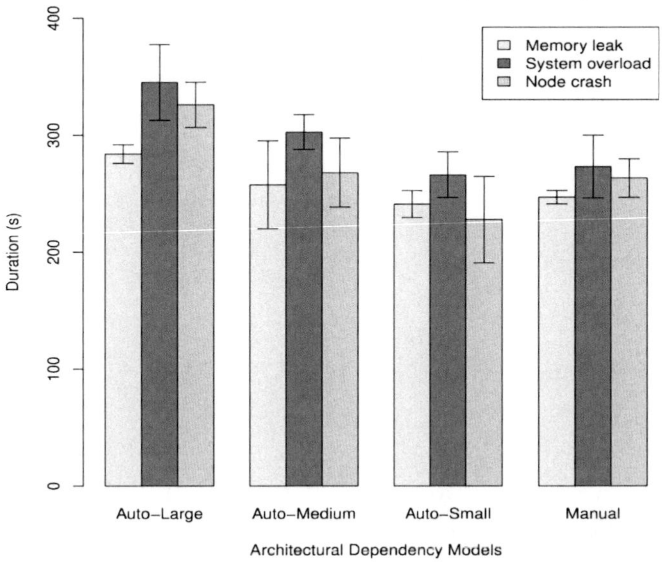

Figure 12: HORA's average analysis and prediction duration (including 95% confidence interval) for a 2 h monitoring data.

On average, the analysis for a 2 h monitoring data is completed in less than six minutes. This demonstrates that HORA can be deployed and make timely predictions at runtime. It needs to be emphasized that HORA's prediction process is triggered by every new data point. For the model with 98 components, this leads to 98 predictions every 2 min, i.e., 5880 predictions in two hours. In this work, we focus on investigating the prediction quality of HORA rather than its prediction efficiency. Particularly, HORA has not been optimized for performance. A possible future optimization could be to configure it to make predictions at regular time intervals, e.g., every 1 minute.

Discussion

Our prediction approach exploits the knowledge of the component dependencies and a set of predictors, which can predict individual component failures, to infer the failure propagation. Our results show that in the memory leak and system overload scenarios, HORA can predict the failures with high TPR. This demonstrates that the problems that develop internally can be detected early and the failure probability can be propagated to other parts of the system.

Although the results in Fig. 6 and Table 4 show that HORA achieves higher TPR and higher AUC, the number of false positive is also high. This results in a low precision and high false-positive rate. In other words, the monolithic approach performs better in the low false-positive-rate range between 0 and 0.1. On the other hand, if a higher false-positive rate is acceptable, HORA will be able to correctly predict more failures than the monolithic approach.

Threats to validity

We inject faults that trigger application failures in our experiments, which is a common practice in accessing dependability (Irrera and Vieira, 2014; Natella et al., 2016), e.g., fault tolerance or failure prediction. It is possible that the failures that occur at runtime may be caused by other hidden problems rather than those that we inject. In our evaluation, the failures that occur in the memory leak scenario can be caused by a system overload if the workload is too high. As a result, an attempt to predict failures caused by a memory leak will also predict failures of the system overload problem. Therefore, we carefully choose the workload that is low enough so that it does not cause system overload while we inject other types of faults.

In order to evaluate our approach, we need datasets that include architectural dependencies, detailed runtime measurements for the architectural entities, and information regarding the types and time of the failures. Publicly available datasets exist,[3] but they are not appropriate for our evaluation because they lack the architectural information and runtime measurements. To systematically evaluate our approach, a controlled environment is needed, which includes a usage profile and the types of failures. We conduct a lab study with fault injection which presents two main threats to external validity. First, we consider only one system. Therefore, we select an open-source application that is representative for the state-of-the-art enterprise systems, in terms of architectural style (microservice-based (Newman, 2015)) and technology (NetflixOSS ecosystem). Second, our experiment did not cover all possible types of faults. Since covering all possible fault types is practically infeasible, we select three representative fault types from real world incidents based on Pertet and Narasimhan (2005). The possibility to reduce these threats for future studies would be that the community makes suitable data available and develop a benchmark for online failure prediction techniques. For this paper, we provide both the system setup and the data so that researchers can replicate our experiment and build on our research.

In the evaluation, we compare the prediction results of the Hora approach with those of the monolithic approach, which employs ARIMA as predictors and does not consider the architecture of the system. There are existing works for predicting component failures. However, they are specifically applicable for different types of monitoring data or different types of components. For instance, Fronza et al. (2013), Liang et al. (2007) and Salfner and Malek (2007) employ machine learning techniques to analyze event logs and classify the system into healthy and failure states. These techniques are not directly applicable as the monitoring data obtained in our experiment are time series data. Therefore, we utilize ARIMA which is a suitable and commonly used prediction technique for this type of data. The relation to other prediction techniques is discussed in detail in the following related work section.

RELATED WORK

In this paper, we propose a novel approach for hierarchical online failure prediction. This work is related to QoS and failure prediction, and can be categorized based on two dimensions; 1) online vs. offline, and 2) monolithic vs. hierarchical.

1. Online prediction approaches aim at providing information regarding the near future state of the running system based on runtime observations (Salfner et al., 2010). In contrast, offline prediction approaches are not used to trigger runtime actions, but to focus on providing QoS measures to reason on system design and evolution decisions (Cortellessa et al., 2011; Musa, 1998).
2. Monolithic prediction approaches consider the system as a black box, of which the architectural information is not known. A prediction model can be created using different techniques, such as time series forecasting or machine learning. On the other hand, hierarchical prediction approaches consider the architecture of software systems including the components and their inter-dependencies. Each component has its own specification that can be combined with the others' to form a model that represents the whole system. The relevant measures of the system can then be obtained by solving the combined model.

The remainder of this section describes related work in three categories, based on the two dimensions, discussed previously. Due to the lack of relevance to our approach, we do not discuss works on monolithic offline prediction.

Monolithic online prediction

Similar to our approach, Cavallo et al. (2010) and Amin et al. (2012a, 2012b) use ARIMA and GARCH models, respectively, to forecast response times and time between failures of web services. In contrast, other approaches use statistical analysis with adaptive thresholds (Bovenzi et al., 2011), complex event processing (Baldoni et al., 2012), rule-based classification (Sahoo et al., 2003), linear regression and decision trees (Alonso et al., 2009), support vector machine (Liang et al.,

2007; Fronza et al., 2013), nearest neighbor methods (Liang et al., 2007), Bayesian networks and submodels (Guan et al., 2012; Watanabe et al., 2012; Fu and Xu, 2007), and hidden semi-Markov models (Salfner and Malek, 2007).

Bovenzi et al. (2011) apply statistical analysis with adaptive threshold to predict threshold violation of performance metrics in complex software systems. Michlmayr et al. (2009) present a framework to predict SLA violations in web services, which combines the advantages of both client- and server-side QoS monitoring.

For black-box failure prediction, Baldoni et al. (2012) employ complex event processing and hidden Markov models to predict failures of a distributed system based on the sniffed network traffic. On the other hand, Williams et al. (2007) monitor performance metrics including response time and predict failures using anomaly detection in combination with dispersion frame technique.

Regarding failures in large-scale systems, Sahoo et al. (2003) predict critical events in computer cluster by analyzing event logs using rule-based classification algorithms. Zheng et al. (2010) predict failures in Blue Gene/P supercomputers by analyzing the log messages and using genetic algorithms to create rules for the log patterns that precede the failures. Liang et al. (2007) predict failures based on the event log from Blue Gene/L using different machine learning techniques, i.e., RIPPER, support vector machine, and nearest neighbor method. Guan et al. (2012) employ Bayesian submodels to detect anomalies in the performance metrics of cloud computing systems. Watanabe et al. (2012) collect log files of a cloud data center and analyze the similarity between the records. The Bayesian method is then used to compute the probability of a failure following certain error sequences. In our previous work (Pitakrat et al., 2014a), we apply machine learning techniques to analyze the correlation of event log of HPC systems and use the model to predict failure events. Fu and Xu (2007) employ Bayesian networks to predict failure events in HPC systems. The failure information is forwarded to a master node in order to adjust the job allocation.

For other types of systems, Salfner and Malek (2007) propose an online failure prediction approach using hidden semi-Markov models. These models utilize both time and type of error messages to predict failures in a telecommunication system. Alonso et al. (2009) predict web server crashes by monitoring the memory usage and applying linear regression and decision trees to predict the time to memory exhaustion. Fronza et al. (2013) apply random indexing to log messages and use support vector machine to predict the patterns that can lead to software failures.

The HORA approach differs from these monolithic approaches at the system level in that we explicitly take architectural information into account. Although we apply similar techniques to predict individual component failures, the architectural knowledge is employed to predict how a failure can propagate and affect other components.

Hierarchical offline prediction

The approaches in this category employ architecture-based system models annotated with specific quality evaluation models or scenarios (Babar and Gorton, 2004; Babar et al., 2004; Grunske, 2007), e.g., performance (Balsamo et al., 2004; Koziolek, 2010), reliability (Goseva-Popstojanova and Trivedi, 2001), and safety (Grunske and Han, 2008) attributes. The model can be solved by using analytical solution or simulations to obtain the relevant properties of the whole system.

Cheung proposes a seminal software reliability model that takes into account reliability of individual components along with the probability of calling other components. A Markov model is employed to combine the reliability of components and represent the reliability of the whole system. Cortellessa and Grassi (2007)present an approach for reliability analysis of component-based software systems. Based on the system architecture, they consider the error propagation probability between components in addition to the reliability of individual components. Becker et al. (2009) introduce the Palladio Component Model (PCM) which enables performance prediction of component-based software systems. Brosch et al. (2012)extend the PCM by annotating the components with corresponding reliability attributes. The model is transformed into a discrete-time Markov chain and solved to obtain the reliability of the system.

The approaches in this category have shown that the capability of the prediction can be significantly improved by combining traditional approaches, which are monolithic, with the architecture-based models. Our approach applies the same concept and incorporates the architecture of the system, specifically the dependencies between components, into online failure prediction. In combination with monolithic prediction approaches, our approach is able to predict both failures of individual components and the probabilities that those failures will propagate to other parts of the system.

Hierarchical online prediction

As one recent example for performance, Brosig et al. (2014) employ an architecture-based performance model to predict system performance at runtime for capacity planning and online resource provisioning. The performance characteristics are captured in an architectural performance model which is then solved by transforming it to an analytical model or by simulation, similar to Becker et al. (2009). As opposed to this, HORA focuses on predicting failure occurrences using an extensible set of tailored prediction techniques.

Chalermarrewong et al. (2012) predict system unavailability in data centers using a set of component predictors and fault tree analysis. The component predictors employ autoregressive moving average (ARMA) to predict failures of hardware components. These component failures are leaf nodes in the fault tree which is evaluated to conclude whether the current set of component failures will lead to system unavailability. Even though this work does not consider software, it shares the same basic idea as HORA by having a dedicated failure predictor for each component. However, the fault tree does not incorporate the conditional probability which represents complex software architectural relationships. On the other hand, HORAemploys Bayesian networks which can represent conditional dependencies and infer the probabilities of failures and their propagation.

CONCLUSION

Failures in software systems usually develop inside the system and propagate to the boundary. Existing online failure prediction approaches do not explicitly consider the software system architecture and failure propagation paths. They view the system as monolithic and make predictions based only on the available measurements, such as response time.

In this paper, we introduce our hierarchical online failure prediction approach, HORA, which employs a combination of a Failure Propagation Model and component failure prediction techniques. The Failure Propagation Model uses Bayesian networks and is extracted from an architectural dependency model. The component failure predictors are updated by continuous measurements of the running system. Our evaluation shows that HORA provides a significantly higher prediction quality than the monolithic approach. In addition to the improved prediction quality, HORA allows a higher degree of modularity as different failure prediction techniques can be applied and reused among similar types of system components. Furthermore, the automatic extraction allows the architectural dependency model to be generated from an existing architectural model. However, HORA can only be used if detailed monitoring data is continuously collected at runtime. Nonetheless, detailed application performance monitoring has become a common practice in large scale systems, providing the data needed by HORA.

In our future work, we plan to investigate and improve HORA's suitability for rapidly changing systems. For instance, emerging software engineering paradigms, such as DevOps (Bass et al., 2015), aim for faster release cycles and rely on dynamic cloud environments. The structures and behaviors of these systems change frequently which requires continuous relearning of the failure models in order to retain the prediction quality (Irrera et al., 2014). We expect that HORA's modular failure prediction approach is suitable under these conditions, because each predictor can be retrained, independently from other predictors. However, architectural changes can alter the failure propagation path. Thus, the Failure Propagation Model needs to be kept up-to-date. Moreover, we will also optimize the predictor to minimize the number of false positives and we plan to extend our evaluation setup to a benchmark for online failure prediction approaches.

ACKNOWLEDGMENT

This work has been partly funded by the German Federal Ministry of Education and Research (grant no. 01IS15004).

REFERENCES

1. Alonso, J. , Torres, J. , Gavalda, R. , 2009. Predicting web server crashes: A case study in comparing prediction algorithms. In: Proc. 5th Int. Conf. on Autonomic and Autonomous Systems (ICAS '09), pp. 264–269 .
2. Amin, A . , Colman, A . , Grunske, L. , 2012a. An approach to forecasting QoS attributes of web services based on ARIMA and GARCH models. In: Proc. 19th Int. Conf. on Web Services (ICWS '12), pp. 74–81 .
3. Amin, A. , Grunske, L. , Colman, A. , 2012b. An automated Approach to Forecasting QoS Attributes based on linear and non-linear Time Series Modeling. In: Proc. 27th IEEE/ACM Int. Conf. on Automated Software Engineering (ASE '12), pp. 130–139 .
4. Avižienis, A. , Laprie, J.-C. , Randell, B. , Landwehr, C. , 2004. Basic concepts and taxon- omy of dependable and secure computing. IEEE Trans. Dependable Secure Com- put. 1 (1), 11–33 .
5. Babar, M.A. , Gorton, I. , 2004. Comparison of scenario-based software architec- ture evaluation methods. In: Proc. 11th Asia-Pacific Software Engineering Conf. (APSEC '04), pp. 600–607 .
6. Babar, M.A. , Zhu, L. , Jeffery, D.R. , 2004. A framework for classifying and compar- ing software architecture evaluation methods. In: Australian Soft. Eng. Conf. (ASWEC 2004). IEEE Computer Society, pp. 309–319 .
7. Baldoni, R. , Montanari, L. , Rizzuto, M. , 2015. On-line failure prediction in safety-crit- ical systems. In: Future Gener. Comp. Syst., 45. North-Holland, pp. 123–132 .
8. Balsamo, S. , Marco, A.D. , Inverardi, P. , Simeoni, M. , 2004. Model-based performance prediction in software development: a survey. IEEE Trans. Software Eng. 30 (5), 295–310 .
9. Bass, L. , Weber, I. , Zhu, L. , 2015. DevOps: A Software Architect's Perspective. Addis- on-Wesley Prof .
10. Becker, S. , Koziolek, H. , Reussner, R. , 2009. The Palladio component model for mod- el-driven performance prediction. J. Syst. Software 82 (1), 3–22 .
11. Bishop, C.M. , 2006. Pattern Recognition and Machine Learning.

Springer .

12. Bovenzi, A. , Brancati, F. , Russo, S. , Bondavalli, A. , 2011. A statistical anomaly-based algorithm for on-line fault detection in complex software critical systems. In: Proc. 30th Int. Conf. on Computer Safety, Reliability, and Security (SAFECOMP '11), pp. 128–142 .
13. Bradley, A.P. , 1997. The use of the area under the ROC curve in the evaluation of machine learning algorithms. Pattern Recognit. 30 (7), 1145–1159 .
14. Brosch, F. , Koziolek, H. , Buhnova, B. , Reussner, R. , 2012. Architecture-based reliability prediction with the palladio component model. IEEE Trans. Software Eng. 38 (6), 1319–1339 .
15. Brosig, F. , Huber, N. , Kounev, S. , 2014. Architecture-level software performance ab- stractions for online performance prediction. Sci. Comput. Program. 90, Part B, 71–92 .
16. Brun, Y. , Bang, J.Y. , Edwards, G. , Medvidovic, N. , 2015. Self-adapting reliability in distributed software systems. IEEE Trans. Software Eng. 41 (8), 764–780 .
17. Calinescu, R. , Grunske, L. , Kwiatkowska, M.Z. , Mirandola, R. , Tamburrelli, G. , 2011. Dynamic QoS management and optimization in service-based systems. IEEE Trans. Software Eng. 37 (3), 387–409 .
18. Candea, G. , Kawamoto, S. , Fujiki, Y. , Friedman, G. , Fox, A. , 2004. Microreboot—a tech- nique for cheap recovery. In: Proc. Symposium on Operating Systems Design & Implementation (OSDI '04), pp. 31–44 .
19. Cavallo, B. , Penta, M.D. , Canfora, G. , 2010. An empirical comparison of methods to support QoS-aware service selection. In: Proc. 2nd Int. Workshop on Principles of Engineering Service-Oriented Systems (PESOS '10). ACM, pp. 64–70 .
20. Chalermarrewong, T. , Achalakul, T. , See, S. , 2012. Failure prediction of data centers using time series and fault tree analysis. In: Proc. 18th Int. Conf. on Parallel and Distributed Systems (ICPADS '12), pp. 794–799 .
21. Cheung, R.C. , 1980. A user-oriented software reliability model. IEEE Trans. Software Eng. 6 (2), 118–125 .

22. Cortellessa, V. , Di Marco, A. , Inverardi, P. , 2011. Model-Based Software Performance Analysis. Springer .
23. Cortellessa, V. , Grassi, V. , 2007. A modeling approach to analyze the impact of error propagation on reliability of component-based systems. In: Proc. 10th Int. Conf. on Comp.-Based Soft. Eng. (CBSE '07), pp. 140–156 .
24. DeLong, E.R. , DeLong, D.M. , Clarke-Pearson, D.L. , 1988. Comparing the areas under two or more correlated receiver operating characteristic curves: a nonparamet- ric approach. Biometrics 44 (3), 837–845 .
25. Fawcett, T. , 2006. An introduction to ROC analysis. Pattern Recognit. Lett. 27 (8), 861–874 .
26. Fowler, M. , 2002. Patterns of Enterprise Application Architecture. Addison Wesley .
27. Fronza, I. , Sillitti, A. , Succi, G. , Terho, M. , Vlasenko, J. , 2013. Failure prediction based on log files using random indexing and support vector machines. J. Syst. Soft- ware 86 (1), 2–11 .
28. Fu, S. , Xu, C.-Z. , 2007. Exploring event correlation for failure prediction in coalitions of clusters. In: Proc. 2 \ 007 Conf. on Supercomputing (SC '07), pp. 41:1–41:12 .
29. Goseva-Popstojanova, K. , Trivedi, K.S. , 2001. Architecture-based approach to reliabil- ity assessment of software systems. Perf. Eval. 45 (2–3), 179–204 .
30. Grunske, L. , 2007. Early quality prediction of component-based systems - a generic framework. J. Syst. Software 80 (5), 678–686 .
31. Grunske, L. , Han, J. , 2008. A comparative study into architecture-based safety evalu- ation methodologies using AADL's error annex and failure propagation models. In: Proc. 11th IEEE High Assurance Systems Engineering Symposium (HASE) '08, pp. 283–292 .
32. Guan, Q. , Zhang, Z. , Fu, S. , 2012. Ensemble of Bayesian predictors and decision trees for proactive failure management in cloud computing systems. J. Commun. 7 (1), 52–61 .
33. Halili, E. , 2008. Apache JMeter. Packt Publishing .
34. Hibler, M. , Ricci, R. , Stoller, L. , Duerig, J. , Guruprasad, S. , Stack, T. , Webb, K. , Lep- reau, J. , 2008. Large-scale virtualization

in the Emulab network testbed.. In: USENIX Annual Technical Conference, pp. 113–128 .

35. Hiller, M. , Jhumka, A. , Suri, N. , 2001. An approach for analysing the propagation of data errors in software. In: Proc. 2001 Int. Conf. on Dependable Systems and Networks (DSN '01), pp. 161–172 .
36. van Hoorn, A. , 2014. Model-Driven Online Capacity Management for Componen- t-Based Software Systems . Faculty of Engineering, Kiel University.
37. van Hoorn, A. , Waller, J. , Hasselbring, W. , 2012. Kieker: A framework for application performance monitoring and dynamic software analysis. In: Proc. Int. Conf. on Performance Eng. (ICPE '12), pp. 247–248 .
38. Huang, J. , Ling, C. , 2005. Using AUC and accuracy in evaluating learning algorithms. IEEE Trans. Knowl. Data Eng. 17 (3), 299–310 .
39. Irrera, I. , Duraes, J. , Vieira, M. , 2014. On the need for training failure prediction algo- rithms in evolving software systems. In: Proceedings of the 2014 IEEE 15th In- ternational Symposium on High-Assurance Systems Engineering. IEEE Computer Society, Washington, DC, USA, pp. 216–223 .
40. Irrera, I. , Vieira, M. , 2014. A practical approach for generating failure data for as- sessing and comparing failure prediction algorithms. In: Dependable Computing (PRDC), 2014 IEEE 20th Pacific Rim International Symposium on, pp. 86–95 .
41. Johansson, A. , Suri, N. , 2005. Error propagation profiling of operating systems. In: Proc. 2015 Int. Conf. on Dependable Systems and Networks (DSN '05), pp. 86–95 .
42. Koziolek, H. , 2010. Performance evaluation of component-based software systems: a survey. Perform. Eval. 67 (8), 634–658 .
43. Kutschke, M., Jayes: Bayesian network library, 2013. https://github.com/kutschkem/ Jayes .
44. Li, Y. , Lan, Z. , 2006. Exploit failure prediction for adaptive fault-tolerance in cluster computing. In: Proc. 6th Int. Symposium on Computing and the Grid (CCGRID '06), pp. 8pp.–538 .
45. Liang, Y. , Zhang, Y. , Xiong, H. , Sahoo, R.K. , 2007. Failure

prediction in IBM BlueGene/L Event Logs. In: Proc. Int. Conf. on Data Mining (ICDM '07), pp. 583–588 .

46. Lorenzoli, D. , Mariani, L. , Pezzè, M. , 2008. Automatic generation of software behav- ioral models. In: Proc. 30th Int. Conf. on Software Engineering (ICSE '08). ACM, pp. 501–510 .
47. Michlmayr, A. , Rosenberg, F. , Leitner, P. , Dustdar, S. , 2009. Comprehensive QoS mon- itoring of web services and event-based SLA violation detection. In: Proc. 4th Int. Workshop on Middleware for Service Oriented Computing (MWSOC '09), pp. 1–6 .
48. Montgomery, D.C. , Runger, G.C. , Hubele, N.F. , 2009. Engineering Statistics. John Wi- ley & Sons .
49. Musa, J.D. , 1998. Software Reliability Engineering. McGraw-Hill .
50. Natella, R. , Cotroneo, D. , Madeira, H.S. , 2016. Assessing dependability with software fault injection: a survey. ACM Comput. Surv. 48 (3), 4 4:1–4 4:55 .
51. Newman, S. , 2015. Building Microservices. O'Reilly Media, Inc .
52. Nygard, M. , 2007. Release It!: Design and Deploy Production-Ready Software. Prag- matic Bookshelf .
53. Pertet, S. , Narasimhan, P. , 2005. Causes of failure in web applications. Parallel Data Lab. 48 .
54. Pitakrat, T. , Grunert, J. , Kabierschke, O. , Keller, F. , van Hoorn, A. , 2014a. A framework for system event classification and prediction by means of machine learning. In: Proc. 8th Int. Conf. on Performance Evaluation Methodologies and Tools (VAL- UETOOLS '14) .
55. Pitakrat, T. , van Hoorn, A. , Grunske, L. , 2013. A comparison of machine learning al- gorithms for proactive hard disk drive failure detection. In: Proc. 4th Int. Conf. on Architecting Critical Systems (ISARCS '13). ACM, pp. 1–10 .
56. Pitakrat, T. , van Hoorn, A. , Grunske, L. , 2014b. Increasing dependability of compo- nent-based software systems by online failure prediction. In: Proc. Euro. De- pendable Computing Conf. (EDCC '14), pp. 66–69 .
57. Pitakrat, T. , Okanovic, D. , van Hoorn, A. , Grunske, L. , 2016. An architecture-aware ap- proach to hierarchical online failure prediction. In: Proc. 12th Int. ACM SIGSOFT Conference on the

Quality of Software Architectures (QoSA '16). IEEE, pp. 60–69 . R Core Team, 2015. R: A Language and Environment for Statistical Computing. R Foundation for Statistical Computing. Vienna, Austria. http://www.R-project. org/ .

58. Robin, X. , Turck, N. , Hainard, A. , Tiberti, N. , Lisacek, F. , Sanchez, J.-C. , Müller, M. , 2011. pROC: an open-source package for R and S+ to analyze and compare ROC curves. BMC Bioinf. 12 (1), 77 .
59. Sahoo, R.K. , Oliner, A.J. , Rish, I. , Gupta, M. , Moreira, J.E. , Ma, S. , Vilalta, R. , Siva- subramaniam, A. , 2003. Critical event prediction for proactive management in large-scale computer clusters. In: Proc. 9th Int. Conf. on Knowledge Discovery and Data Mining (KDD '03), pp. 426–435 .
60. Salfner, F. , Lenk, M. , Malek, M. , 2010. A survey of online failure prediction methods. ACM Comput. Surv. 42 (3), 10:1–10:42 .
61. Salfner, F. , Malek, M. , 2007. Using hidden semi-Markov models for effective online failure prediction. In: Proc. 26th Int. Symposium on Reliable Distributed Systems (SRDS '07), pp. 161–174 .
62. Schmerl, B.R. , Aldrich, J. , Garlan, D. , Kazman, R. , Yan, H. , 2006. Discovering architec- tures from running systems. IEEE Trans. Software Eng. 32 (7), 454–466 .
63. Shumway, R.H. , Stoffer, D.S. , 2013. Time Series Analysis and its Applications. Springer Science .
64. Walker, R. , 2006. Examining load average. Linux Journal 2006 (152), 5–16 .
65. Watanabe, Y. , Otsuka, H. , Sonoda, M. , Kikuchi, S. , Matsumoto, Y. , 2012. Online fail- ure prediction in cloud datacenters by real-time message pattern learning. In: Proc. 4th Int. Conf. on Cloud Computing Technology and Science (CloudCom '12), pp. 504–511 .
66. Williams, A.W. , Pertet, S.M. , Narasimhan, P. , 2007. Tiresias: Black-box failure pre- diction in distributed systems. In: Proc. of Parallel and Distributed Processing Symposium (IPDPS '07), pp. 1–8 .
67. Wohlin, C. , Runeson, P. , Höst, M. , Ohlsson, M.C. , Regnell, B. , Wesslén, A. , 2012. Ex- perimentation in Software Engineering. Springer Science & Business Media .

68. Zheng, Z. , Lan, Z. , Gupta, R. , Coghlan, S. , Beckman, P. , 2010. A practical failure pre- diction with location and lead time for Blue Gene/P. In: Proc. 2010 Int. Conf. on Dependable Systems and Networks Workshops (DSN-W '10), pp. 15–22

CHAPTER 10

CONTINUOUS DELIVERY: OVERCOMING ADOPTION CHALLENGES

Lianping Chen

Lianping Chen Limited, Dublin, Ireland

ABSTRACT

Continuous Delivery (CD) is a relatively new software development approach. Companies that have adopted CD have reported significant benefits. Motivated by these benefits, many companies would like to adopt CD. However, adopting CD can be very challenging for a number of reasons, such as obtaining buy-in from a wide range of stakeholders whose goals may seemingly be different from—or even conflict with—our own; gaining sustained support in a dynamic complex enterprise environment; maintaining an application development team's momentum when their application's migration to CD requires an additional strenuous effort over a long period of time; and so on. To help overcome the adoption challenges, I present six strategies: (1) selling CD as a painkiller; (2) establishing a dedicated team with multi-disciplinary members; (3) continuous delivery of continuous delivery; (4) starting with the easy but important applications; (5) visual CD pipeline skeleton; (6) expert drop. These strategies were derived from four years of experience in implementing CD at a multi-billion-euro company. Additionally, our experience led to the identification of eight further challenges for research. The information contributes toward building a body of knowledge for CD adoption.

Keywords

Agile Software Development, Continuous Delivery, Continuous Deployment, Continuous Software Engineering, DevOps, Adoption

INTRODUCTION

Continuous Delivery (CD) is a software engineering approach in which teams keep producing valuable software in short cycles and ensure that the software can be reliably released at any time (Chen, 2015a).

The CD approach is relatively new. It started gaining wide attention only in 2010, when Humble and Farley published the book titled "Continuous Delivery: Reliable Software Releases through Build, Test, and Deployment Automation" (Humble and Farley, 2010). However, the CD approach has become increasingly popular, as shown by Google search trends (Fig. 1).

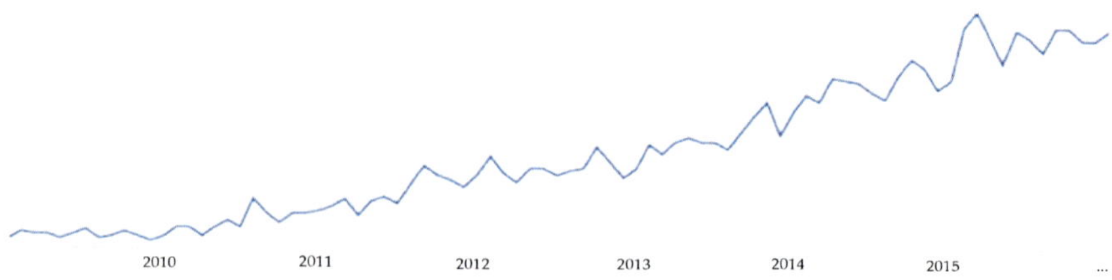

Figure 1: Five-year Google trend for the search term "Continuous Delivery".

Companies that practice CD have reported huge benefits, such as significant improvements in time-to-market, customer satisfaction, product quality, release reliability, productivity and efficiency, and the ability to build the right product through rapid experiments (Chen, 2015a; Leppanen et al., 2015).

These benefits have motivated many companies to adopt CD. According to a recent survey of 600 software developers, managers, and executives in the United States and the United Kingdom, only 3% of the respondents said they had no plans to adopt CD (Perforce Software Inc., 2015).

However, implementing CD can be quite challenging (Chen, 2015a; Leppanen et al., 2015; Claps et al., 2015). Although CD as a goal (a target state) is no longer a new idea and has been well documented (Humble and Farley, 2010), the adoption journey for CD is not yet a smooth path. The journey itself is where the challenges lie, where many companies struggle, and where practitioners need help. Some of the fundamental challenges include the following:

- acquiring buy-in from a wide range of stakeholders whose goals may seemingly be different from—or even conflict with—those of the team driving the CD implementation;
- gaining sustained support in a dynamic complex enterprise environment;
- Maintaining an application development team's momentum when their application's migration to CD requires an additional strenuous effort over a long period of time.

To help overcome the adoption challenges, I present six strategies we learned along our journey to CD at a multi-billion-euro company called Paddy Power. Paddy Power has been in the process of implementing CD for the past four years and, consequently, we have encountered and overcome many challenges. However, eventually, we achieved huge benefits (Chen, 2015a).

I hope the strategies described here can help fellow practitioners to overcome similar challenges on their way to achieving CD. The information also contributes toward building a body of knowledge for CD adoption.

I also outline eight further challenges for research. Although not all of them are new, I provide elaboration on why these challenges are important for the industry, which on its own right can be useful input to researchers.

The remainder of this paper is organized as follows: Section 2 describes the context from which the strategies described here were derived. Section 3 describes the six strategies. I discuss related work in Section 4 and outline further research challenges in Section 5. Section 6 summarizes the paper.

THE CONTEXT

Paddy Power

Paddy Power is a rapidly growing company in the bookmaking industry. It offers its services in regulated markets, through betting shops, phones, and the Internet. It has an annual turnover of approximately €7 billion and employs approximately 5000 people. Paddy Power recently merged with Betfair, making it the world's largest public online betting and gaming company.

The company relies heavily on an increasingly large number of custom software applications that include websites, mobile apps, trading and pricing systems, live-betting-data distribution systems, and software used in betting shops. We developed these applications using a wide range of technology stacks, including Java, Ruby, PHP, and .NET. To run the applications, the company has an IT infrastructure consisting of thousands of servers in different locations. The applications are developed and maintained by the Technology Department, which employs approximately 500 people.

Continuous Delivery

At Paddy Power, we view Continuous Delivery as a software engineering approach in which teams keep producing valuable software in short cycles and ensure that the software can be reliably released at any time (Chen, 2015a). The following sections highlight the characteristics of CD that are important to us.

Valuable software

Developing valuable software is a goal that has long been on the Agile manifesto (Beck et al., 2001). However, it is not an easy goal to achieve. Before adopting CD, some of our teams had been using an Agile method called Kanban (Anderson, 2010); however, due to delivery problems, we still had situations where a team had completed a feature but could not deliver it to production to obtain users' feedback. Consequently, they built additional functionalities on top of that feature, simply assuming it was useful. Unfortunately, when they finally delivered the software to the users, they found out that the feature was not what the users needed. Even worse, by that point, significant effort had been spent on the feature and the additional functionalities. An important objective of our CD implementation was to alleviate this problem. We want teams to build valuable software rather than spend time on features that users do not need.

Short cycles

We use "cycle time" to refer to the time from the conception of a user story to its production. Our cycle time used to be multi-month. Because of this long cycle time, the user stories that were completed earlier in the cycle had to wait for a long time. The value they could otherwise have generated was lost. Moreover, we were unable to obtain early user feedback. Therefore, shortening the cycle time is important.

Releasable at any time

Before implementing CD, applications were only releasable at the end of a long release cycle. This constraint caused problems. For example, business people could not obtain a release to users on demand in the middle of this long cycle, no matter how desperately they needed it; users were simply forced to wait until the next big release to obtain some important features—although those features might have been developed early, at the beginning of the cycle. To eliminate these problems, we want applications to be releasable at any time.

Reliable releases

Our emphasis on reliable releases stem directly from bad release experiences. Each time we were about to release, we had little confidence in the release reliability. Many times, these releases would be followed by P1 (priority 1) incidents (Rob, 2007), meaning that release activity was always full of uncertainty, failures, and stress. Reliable releases are a big point of distinction between our old practices and CD.

A term very similar to CD exists: Continuous Deployment. In academia, people tend to use it interchangeably with Continuous Delivery (Rodríguez et al., 2016).

However, many practitioners tend to clearly distinguish these two terms. The distinction is that under Continuous Deployment we deploy any change to production that passes a series of tests. In contrast, under Continuous Delivery, we ensure that the software can be reliably released at any time, but it is up to a human to decide when to release. In both Continuous Deployment and Continuous Delivery, deployment to production itself is automated. The difference lies in the trigger for making the deployment. One is triggered automatically, while the other is triggered by a human. According to this distinction, Continuous Delivery is compatible with a wide range of scenarios, but Continuous Deployment is suitable only under special conditions (O'Dell and Skelton, 2016). As do most companies, we mainly use Continuous Delivery. However, due to their similarities, the strategies reported in this paper likely also apply to the adoption of Continuous Deployment.

CD at Paddy Power

Paddy Power's journey to CD started four years ago, led by a dedicated team of eight people (the CD team). This team is primarily responsible for building the Continuous Delivery Platform (our tool chain to support continuous delivery), guiding and supporting the development teams in moving their applications to continuous delivery. Due to the considerable diversity of the teams, the guiding and supporting activities vary significantly from team to team. The author is part of this CD team. The CD team controls neither the development teams nor their applications; thus, it can provide only guidance and support and leave the development teams themselves to actually make the necessary changes to their development practices and applications. The changes that a development team needs to make also vary significantly from team to team. Some example changes include increasing code coverage by implementing automated tests so that the development team can be confident that when a code change passes all the stages of the CD pipeline it is ready to release; modifying an application to increase its testability so that automated tests can be written with an acceptable level of cost; increasing an application's modifiability, deploy ability, etc. as discussed in Chen (2015b); breaking big features down into smaller ones so that each can be finished in a short time (e.g., a week).

At present, we have implemented CD for 60 different applications. For each of these applications, a developer making a code commit automatically triggers a CD pipeline execution. The updated code undergoes a series of stages that check the quality of the code change. If the change fails to pass at any of the pipeline stages, the pipeline is aborted, and the developers are notified so they can fix the problems. If the updated code passes all the stages, it can be released to production with the click of a button (Chen, 2015a).

As a result of the CD implementation, we have achieved huge benefits, as summarized in Table 1 (these were reported in an earlier paper (Chen, 2015a)).

Table 1: Summary of benefits achieved in our CD adoption.

Benefits	Description
Accelerated time to market	The cycle time from a user story's conception to production has decreased from several months to two to five days. Release frequency increased from once every one to six months to once a week.
Building the right product	Frequent releases let the application development teams obtain user feedback more quickly. This lets them work on only the useful features. If they find that a feature is not useful, they spend no further effort on it. This helps them build the right product.

Improved productivity and efficiency	Developers used to spend 20% of their time setting up and fixing their test environments. Now, the CD pipeline automatically sets up the environments. Operations engineers used to expend a few days' effort to release an application to production. Now, it can be done with a developer's click of a button.
Reliable releases	Releases have become reliable. The high level of stress and uncertainty associated with release activities has gone. Some teams even make releases at peak times, which they could never have considered before.
Improved product quality	The number of open bugs for applications has decreased by more than 90%. Previously, approximately 30% of the workforce was fixing bugs. Now, usually nobody is working on customer found bugs. Bugs are so rare that the teams no longer need a sophisticated bug-tracking system.
Improved customer satisfaction	Previously, distrust and tension existed between the users' department and the software development teams, owing to quality and release issues. The managers commented that the relationship has improved enormously.

However, our journey to CD was not smooth. We encountered many challenges along the way. For example, implementing CD requires support from different stakeholders, but obtaining buy-in from them is quite difficult. CD implementation requires a considerable and continuous investment over a long period of time (Savor et al., 2016), but gaining sustained support in a dynamic complex enterprise environment is challenging. Moving complex applications to CD can take months or even years because of the number of changes that need to be made to the team's mindset and practices as well as to the application itself. Maintaining the team's momentum toward CD adoption under such circumstances is challenging.

When we were working through these fundamental challenges, we looked to the research literature for possible solutions. Unfortunately, we were unable to find information that specifically addressed how to overcome these CD adoption challenges. It would have been helpful if such guidance had been available because it could have saved us a great deal of time and prevented us from taking wrong turns during our exploration and quest for the right solutions.

I think other fellow practitioners that are in the process or plan to implement CD may have similar needs and will likely encounter similar challenges. Therefore, I decided to present the strategies we learned along our path to implementing CD. In the next section, I will describe each of the six strategies. Henceforward, this paper refers to these strategies as S1 through S6.

STRATEGIES TO OVERCOME CD ADOPTION CHALLENGES

Selling CD as a painkiller (S1)

Adopting CD requires making changes in many areas. Apart from automating build, test, and deployment activities, introducing CD also requires changes to architecture activities (Chen, 2015b), software development practices, organizational structure and culture, and so forth.

To make such substantial changes, we need support from a wide range of stakeholders including top management, the infrastructure team, the system engineering team, the security team, and the operations team.

However, gaining buy-in from all these groups is an enormous challenge because all of them have their own day-to-day work and their own goals and priorities, all of which can seemingly be different from or even conflict with the goal of our team—the team attempting to implement CD.

One strategy that we found useful is selling CD as a painkiller. Using this strategy, we identify each stakeholder's pain points and identify the ones that CD can help to solve. When we introduce CD to this stakeholder, we focus on explaining how and why CD can help to solve the identified pain points.

After the stakeholders realize how CD can help them with their pain points, they often become proponents of CD adoption. I provide two examples below.

An example with software development teams

Software development teams form an important stakeholder in CD implementation. This is because, eventually, it is the software development teams who are responsible for following CD practices and using the CD pipelines. If we do not obtain buy-in from the software development teams, the CD adoption will fail.

It may seem easy to obtain buy-in from the software development teams, but in reality it is not straightforward. For example, our software development teams were often under time pressure. For some teams, overtime work was not uncommon. To squeeze time from their already busy schedules, we needed to give them a very strong motivation.

The first software development department where we implemented CD had approximately 70 employees. A software development team's size depends on the application's size and complexity. The teams range from two to twelve people; most teams have four to eight people. We analyzed the pain points they were experiencing and identified one major pain point that CD could solve that concerned testing environments. In this department, developers on average spent 20% of their time setting up or fixing their testing environments. This was a very boring and frustrating task.

In addition, because the testing environments were shared among teams, a team often needed to wait for one or two weeks for a testing environment to be released by other teams. For new projects requiring new machines, the teams often had to wait for a month before they could obtain the required testing environments from the system engineering and infrastructure teams.

Moreover, the unavailability of testing environments often cause user stories to pile up in the "ready for acceptance" column of the teams' Kanban boards (Anderson, 2010) (Fig. 2), causing further pain.

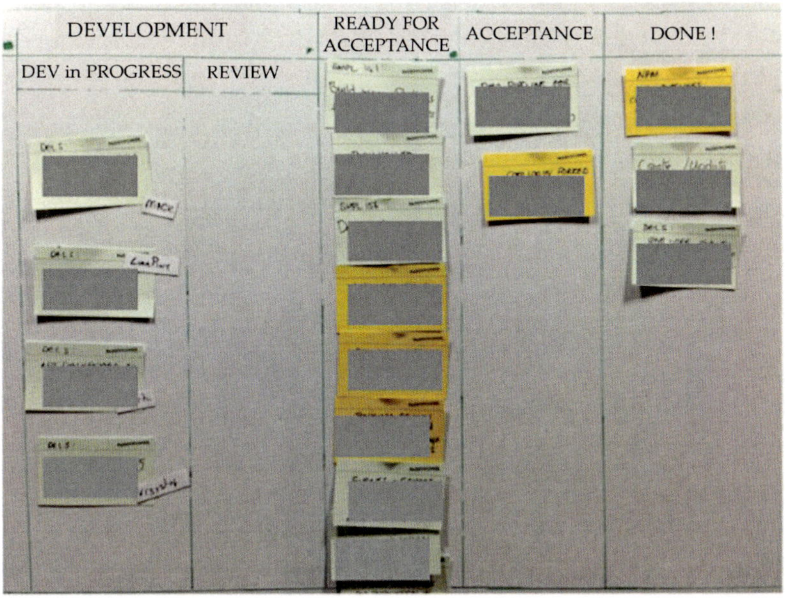

Figure 2: An example of a Kanban board with many stories piled up in the "ready for acceptance" column due to the unavailability of testing environments.

While these stories were waiting in the "ready for acceptance" column, developers would usually pick up another story to work on. Then, when the testing environments finally became available, the developers had to switch back to that story in the "ready for acceptance" column. This context switch incurred extra cost and frustration.

The wait time spent in "ready for acceptance" mode also caused delays because the development team was unable to obtain early user feedback on the waiting stories. Note that the company also loses the value during the delay period that would otherwise be generated if those waiting user stories were released.

CD can eliminate the problems described above. On the CD platform, a development team can create a CD pipeline for each application with support from the CD team. The CD pipeline can automatically spin up all the testing environments within a few minutes. Thus, the developers would no longer need to wait for weeks to obtain testing environments for a new project, nor would they need to spend time setting up and fixing the testing environments.

When the above benefits of CD were explained to the development teams, the developers in that department were keen to move to CD and happy to schedule the time needed to make the changes required on their side.

An example with security team

The security team is an important stakeholder. In our company, before any change can go to production, it must be approved by the security team. The security team's authority is justified. When a security vulnerability goes to production and the system is compromised, the negative effects cannot be undone. We can roll software changes back, but we cannot roll back any security damage that has already occurred. Thus, the company does not allow any change to go to production without the security team's approval.

Without buy-in from security team, we would not be able to implement the crucial step of fast delivery to production. With CD, we needed to release at a much higher frequency, changing from no more than six releases a year to at least one release per week per application. This increased release frequency required more security checks to be conducted. However, considering the large number of applications, the security team's existing practices could not handle the new increased workload. Thus, we needed to make changes—and to make changes in the security area, we needed the security team's buy-in.

We talked to security team and figured out their pain points. CD could solve two of these important pain points. The first pain point concerned ensuring security policy. Although one of these policies was that every change had to go through security checks before it could enter production, because the security scans were performed manually, due to human mistakes, there were situations in which the code that was actually sent to production was not the same code that had been checked by the security team. The following is an example.

A team decided to schedule a release. The source code was tagged in the version control system. The security team checked that code and approved it. However, immediately after the security check, the team made some more changes to the code. The changed code was tagged again in the version control system, and when the release was executed, the code with this new tag (which had not been checked by the security team) was released to production.

The above situation looks like an error that should never happen. However, when the number of applications the security team needs to approve is large (several hundreds) and the process is manual, human errors inevitably happen occasionally.

The security team also had another pain point. There were not enough security engineers to run security checks frequently—for each code commit. For most software applications, the security check was often the last step in the release cycle. When the development team was in a hurry to release code to production (e.g., to adhere to an already slipped deadline), because the security team did not allow them to release until the security checks were finished, developers viewed the security team as an impediment to that release. In such cases, the security team was blamed, which created tension between the security team and the software development teams.

With CD, the execution of security checks is automated. The automation is built into one of the early stages of the CD pipeline. Thus, every code commit goes through the security check. If a security vulnerability is identified, the developers must fix it before they can proceed any further. This ensures that the security policy is always followed.

In addition, the CD pipeline produces an audit trail of the changes made for each code commit. The audit trail includes when the security scan was conducted, what was checked, and what the results of checking were.

This also solves the second pain point. Because the security check is run early and frequently, when the application is to be released, all the security problems have already been identified and fixed. Consequently, the development teams no longer need to wait for the security check, and the security team no longer gets the blame because they no longer stand in the developers' way at the last minute, when the development teams are keen to make the release.

After communicating these advantages to the security team, they became interested in collaborating. Then, after the security team had experienced the benefits from the initial few teams, they became proponents of CD and began to suggest that other teams move to CD as well.

I found that the pain points often differ from one stakeholder to another. Therefore, each stakeholder should be treated individually. I also anticipate that the pain points will differ from one organization to another. So, I do not claim that the examples above are what fellow practitioners will experience or should follow. However, I believe the general strategy is likely to be applicable.

Establishing a dedicated team with multi-disciplinary members (S2)

Establishing a dedicated team with multi-disciplinary members can help to smooth the adoption journey. I will explain why a dedicated team is needed and why the team should contain multi-disciplinary members.

Dedicated team

Implementing CD requires a dedicated team for two reasons. First, implementing CD requires a big effort. Building the CD platform involves a major effort; moving each individual team to CD also requires a significant amount of effort, particularly in a large organization. We have been implementing CD with an eight-person team for four years, amounting to 32 person-years of effort.

Second, without a dedicated team, the CD implementation initiative often suffers from disruptions that make it difficult to make visible progress. For example, before establishing a dedicated CD team, some engineers had been trying to implement CD on a part-time basis. However, no real progress was made, because when the engineers were not dedicated to CD implementation, they were too easily pulled off-task to work on other value streams. Often, for example, the engineers were shifted to work on a late project to help maintain the deadline.

Consequently, we decided to establish a dedicated team. No matter how busy other teams were, the engineers in this team were not switched to other work. Only after this dedicated team was established, did we start to see steady progress in the CD implementation.

Multi-disciplinary members

Implementing CD requires a team with multi-disciplinary members—another lesson that we learned during our journey. When our team was initially established, all the team members had a software development background. However, we soon discovered that developing a CD platform needs not only developer skills but also system administration and operations skills. To implement the CD platform, we would need to bring in many new tools. System administrators are good at installing and maintaining these tools. A large component of the CD platform is the capability to automatically deploy an application, but automating the deployments needs a great deal of operating system configuration, application configuration, and network configuration knowledge. System administrators and operation engineers have deep knowledge in these areas, whereas developers often do not.

Because our team initially had only engineers with a software development background, we were forced to perform these tasks despite the fact that our skills did not match the tasks, which caused the CD adoption to take a longer time than necessary.

Apart from the skills mismatch for performing deployment tasks, communication with the system engineering team (which is responsible for system administration and operations), was not smooth. Because of our developer background we spoke a different language than system engineers. Moreover, we had a lack of appreciation for their concerns, which caused tension between both teams.

To solve this problem, we decided to add a team member who had many years of experience and deep knowledge in system administration and operations. After this member joined our team, we not only accelerated progress in those work areas but also eased the tension with the system engineering team.

When we started to move the more complex applications, whose development process was less amenable to CD, we found that before these teams can effectively practice CD and obtain the benefits of our CD platform, their existing software development practices and processes had to be improved. Expertise in process improvement was very valuable in driving changes in the practices of the development teams, moving them toward the practices required by CD. Having learned from experience, we added a member with process improvement expertise. Now, our team consists of members who have backgrounds in development, system administration, operations, and process improvement. This diversity not only provides us with the required skills to complete work in different areas but also eases communications with other related teams in the organization.

We have gone through a long process to reach our current multi-disciplinary team; therefore, I recommend that any organization planning to adopt CD should consider building a multi-disciplinary team from the beginning of the journey. The exact mixture of skills may vary from one company to another, but at minimum, I suggest including the development and operations disciplines.

Continuous delivery of continuous delivery (S3)

As mentioned above, implementing CD can require a significant effort. We have already spent 32 person-years over four years. To finish something of this size over a long period of time, we need sustained support.

However, getting sustained support in today's dynamic business environment is challenging. Over the last four years, our CEO has changed; our CTO has changed; our manager's manager has changed five times; our organization has been re-structured three times. Whenever such changes occurred, the value of the CD implementation initiative was questioned.

To maintain sustained support and prevent the CD implementation initiative from being killed, we could not afford to deliver the value of CD only after 32 person-years of effort had been spent. We needed to demonstrate the value of CD by delivering it early in the process. Although the total effort so far is 32 person-years, we began onboarding the first application to CD when we were only several months into the CD adoption journey.

With the benefits demonstrated early on, the CD implementation survived many business and organizational changes and maintained sustained support from the company. This support enabled us to develop more features of the CD platform to onboard additional project types. For example, our support for technology stacks has been extended from only

Java to Ruby, PHP, Javascript, Golang (Donovan and Kernighan, 2015), Scala (Odersky et al., 2016), and .NET.

Apart from delivering value to the company early to maintain sustained support, this strategy can also help to reduce the risk and difficulty of CD platform adoption by development teams. We found that when introducing CD, the development teams needed to change the ways in which they typically worked. In other words, the way development teams work needs to be aligned with the new process and the new tools.

Using this strategy, the development teams gradually align their work habits to the new process and new tools in small steps each time we release a new feature of the CD platform. This gradual shift makes the alignment easier.

At the same time, it also provides us with the opportunity to get feedback from the application development teams, for example, to let us know whether a feature we added to the CD platform is useful, easy to use, and so on. We often make adjustments to the CD platform based on developers' feedback. These adjustments align the CD platform to the developers' needs and, thus, simplify its adoption by the development teams.

If we had started to onboard the development teams only after we had completed the entire CD platform, the development teams might have had to make lots of changes to align with the new platform. Large gaps between the way the teams typically working and the CD platform will hinder its adoption.

Thus, continuous delivery of continuous delivery not only helps to address the challenges in maintaining sustained support by the company but also reduces the risk and difficulty that software development teams will have when adopting the CD platform.

Is it possible to decompose the entire CD adoption work into smaller pieces in such a way that when we finish one piece we can gain immediate benefits from it? This is difficult but possible, according to our experience. The following are some dimensions along which one can slice thc CD adoption work into smaller pieces.

Technology Stack: In a large organization, multiple programming languages are used. Each programming language has its own community which typically has developed other tools and technologies to create a full technology stack. The applications developed using different technology stacks use different tools and approaches for compiling, building/packaging, distributing, testing, configuring, deploying, etc. To support all these technology stacks used in a company requires significant effort.

We can build the support for the technology stacks incrementally. When we build the support for one technology stack, we can start onboarding applications using that technology stack to CD, and we can start gaining the benefits of CD from these applications.

Operating System Type: A large organization often uses different operating systems that can be quite different in terms of installation, authentication management, licensing management, scripting environment (e.g., Bash in Linux vs. Powershell in Windows) and so on. To support all these operating systems can require a significant effort.

We can build the support for various operating systems one by one. When we finish building the support for one operating system, we onboard applications running on that operating system, and we can start gaining the benefits of CD from these applications.

Environment (Pipeline Stage): In a large organization, a piece of software needs to be tested in several environments and, finally, be deployed to the production environment. Different environments have different requirements. For example, the requirement for an early-stage testing environment will be different from the requirement for the production environment. Building the supports for automatically provisioning and deploying to all these environments requires a significant effort.

We can build the support for these environments incrementally. When we build the support for one environment, we save the costs for manually setting up and maintaining that environment, which is an immediate benefit to the company.

Degree of Automation: Automating the provisioning of environments can be done in different degrees. A complete provisioning of an environment could include many things such as deploying an application to a virtual machine, creating the virtual machine to which the application needs to be deployed, creating the network that the virtual machine is part of, configuring the network access control list and firewall rules to allow the newly created virtual machine to communicate with other machines, and configuring the different layers of load balancers to make the newly installed application visible to its users. Fully automating all the above requires a significant effort and can take a long time because it not only requires technical changes but also changes to the way that an organization manages infrastructure.

We can start by automating the deployment of applications to pre-provisioned virtual machines. Although the provision of the virtual machines is not automated, we still can begin to benefit from reducing the configuration drifts caused by manual deployments and save the effort of doing manual deployments. We can then proceed to automate virtual machine provision and gain the benefits of not having to create, configure, and maintain the virtual machines manually, finally, we can then automate network provision to gain the corresponding benefits—and so on and so forth.

The above are some heuristics that one can use to decompose the CD adoption work into smaller pieces each of which can be completed in a shorter time, while gaining the benefits from each completed piece immediately. However, one caveat to remember is that we need a clear blueprint to ensure that everything will fit together nicely at the end.

Starting with easy but important applications (S4)

Gaining sustained support is a big challenge, as I already mentioned above, and it is especially challenging in the early stages of the CD implementation. Many people in the organization tend to question CD's effectiveness, its suitability to the organization, the feasibility of its implementation, and whether the CD implementation work has a higher priority than many other work streams that also need people to work on them.

To justify the effort involved in CD implementation, avoid the adoption initiative being killed, and achieve sustained support, we have found another strategy to be quite useful. This strategy involves selecting the first few applications to migrate CD.

Choosing the initial applications is an important topic. In a large organization, both applications and teams can be very diverse. Applications can differ widely in size, complexity, architectural styles, and so forth. The teams differ in size, experience, development practices, culture, their relationships with users, and so on. There are often a large number (hundreds) of applications to select from.

If the wrong application is selected, it may be impossible to demonstrate sufficient CD value to justify the effort and priority required by the CD implementation initiative. Alternatively, selecting the wrong application may make it impossible to complete its migration to CD within the time frame expected by the important stakeholders. In either of these situations, achieving sustained support will become even more difficult.

Based on our experience, the best strategy is to start with the applications that are easy to migrate—but at the same time, to select those that are important to the business. I will explain the rationale below.

Easy to migrate

If the first application that we select to migrate to CD is difficult, it may take too long for us to complete the migration and demonstrate the value of CD. In the worst case, the migration may fail, which—apart from its impact on morale and momentum—will put the entire CD implementation initiative at risk.

Applications that are difficult to migrate to CD do exist in most large organizations. Migrating these applications often requires significant changes to both the application itself and the development team's practices. The changes to the application itself may include such things as architectural changes (Chen, 2015b) and test coverage improvements; the changes to the development team's practices can include changes in architecture practices, testing activities, build processes, and so on. Making these changes can require significant time and effort from the team.

For example, Paddy Power had a large monolithic application. To deploy the application, we needed to take the whole system down. Changing one component often required changes in many other components due to ripple effects. The automated test coverage was poor and many tests were flaky. The quality of the software was mainly ensured by a long round of manual testing at the end of each long release cycle. To move this application to CD, we needed to improve the architecture to satisfy the deployability, modifiability, and testability required by CD

(Chen, 2015b). We also needed to increase the number of automated tests to a degree until the team would be confident about the application's production readiness when it passed the tests.

Making the above changes required huge efforts. This level of effort can impact the pace of delivery of features required by the business stakeholders. Juggling and balancing the CD implementation efforts and the pace of delivery of features is a challenge and sometimes required fighting for the priority of the tasks to make the changes relevant to the CD implementation. Thus, the time required to move this application to CD and showcase CD's benefits could be very long. During this long process, the teams could easily lose the momentum required to make the necessary changes for the move to CD. When we attempted to move this application to CD, progress was slow and maintaining momentum was difficult. We made little progress even over a period of more than 6 months.

If we had selected this application as the first application to move to CD, we would have faced challenges from the management team regarding the benefits and applicability of CD to our organization. The CD implementation initiative might have been killed off at an early stage. Fortunately, we began migrating this application only after having demonstrated the value of CD with several other projects, so we survived through this tough case.

As a heuristic, the applications that are easy to move to CD are often new applications with open-minded teams working on them. New applications do not contain as much legacy things and are more likely to use modern architectural patterns. It is generally easier for teams working on new applications to (re-) design the architecture in a way that is amenable to CD (Chen, 2015b). Moreover, an open-minded team can be easily influenced to adopt the required development practices. For example, we ask new teams to use Kanban, Test Driven Development (TDD) (Beck, 2002), Behavior Driven Development (BDD) (Wynne and Hellesoy, 2012), and other, similar development methodologies.

Important to business

The selected application needs to be important to the company. This helps to secure the required resources, demonstrate clear and unarguable value, and raise the visibility of CD in the organization.

An important application tends to attract resources. Moving an application to CD requires resources, e.g., we need people to spend time to make the necessary changes to the application and implement new development practices. Most organizations have more tasks than resources available to perform them; consequently, it is common to see projects competing for resources. An important application will have a higher probability of acquiring the resources required, whereas unimportant applications will often fail to gain the required resources.

We sometimes need to add additional features to the CD platform to support a new type of application. Such additions can require significant development effort and a sizable investment in infrastructure. For example, to move an application developed using .NET, we needed to make significant changes to our CD platform and make a sizable investment in infrastructure. This is because .NET is quite different from the other applications we already supported. First, .NET applications run on the Windows operating system, and Windows works quite differently from the Linux operating systems we were already supporting in terms of licenses, security, and so on. Thus, we needed to make significant changes to our CD platform. Windows operating systems also require more disk space; therefore, we needed to purchase additional hardware. However, because the application was critical to the business and had top priority, we were able to obtain the required resources.

Equally important, moving an important application to CD tends to generate value that is clearly more than the cost of CD implementation. We have applications that generate millions in revenue monthly for the company. We also have applications that generate tens of thousands in revenue monthly. CD can significantly reduce the time to market and improve product quality (Chen, 2015a). Therefore, moving important applications to CD demonstrates the significant value of CD adoption. Anybody can easily conclude that the value generated is clearly much higher than the cost of CD adoption. While moving less important applications to CD may still generate more value than the cost of CD adoption, the value will be less apparent and less dramatic than moving important applications. Being able to demonstrate clear value generated by CD adoption is important to convince top management to continue to invest in CD adoption. An important application tends to have higher visibility in the organization. So, when this application moves to CD, the success and the value gained from CD adoption will be more visible to the rest of the company and can help drive interest in having more teams move to CD. Thus, migrating important applications helps to create momentum for CD adoption across the whole company. If the application is not important to the company, even if it is successfully moved to CD, people may not care, leaving the CD implementation team as a small hidden corner of the organization.

However, the strategy of starting with easy but important applications does not mean ignoring, avoiding, or not caring about other applications. Instead, it is about onboarding the applications in the right sequence to obtain sustained support for the CD adoption initiative and maximize the value of CD implementation.

After we had moved several applications to CD, the value of CD became concrete and tangible in the company. Motivated by these demonstrated benefits, other teams that were not initially amenable to CD began to show interest in adopting it and became willing to make the changes that they would not have been willing to make several months earlier. By that time, the management team had also already noted the benefits of CD, creating a growing motivation for moving additional applications to CD. This is a better point at which to tackle the tough applications—those that are difficult to move to CD.

Visual CD pipeline skeleton (S5)

Applications that are difficulty to move to CD do exist in most large organizations, and there are often many of them. We have a large number of such tough applications as well, but we cannot simply ignore them in the process of CD adoption. Although it may not make sense to move all these tough applications to CD, many are critical to the company and, therefore, it is worthwhile to migrate them to CD so that new application releases can enter the market faster and provide competitive advantages for the company.

As discussed earlier, some applications are difficult to move to CD because they require substantial changes to both the development team's current practices and to the software applications themselves, such as increasing code coverage by automated tests so that the development team can be confident that when a code change passes all the stages of the

CD pipeline it is ready to release, modifying an application to increase its testability so that automated tests can be written with an acceptable level of cost, and increasing an application's modifiability, deployability, etc. as discussed in Chen (2015b). For these applications, a complete migration to CD can take months or even years to complete.

It is worth noting that our team does not own the applications; we cannot make changes to these applications. Instead, we need each application's development team to make the needed changes. Because we do not have control over the code for these applications, we can only influence and facilitate the improvements.

Under this circumstance, a major challenge is to maintain the application development teams' momentum for CD adoption over a long period of time. We have sometimes seen teams start a migration to CD with strong motivation, but that momentum gradually wanes over time.

One strategy we found useful is to give each team a visual CD pipeline skeleton that shows the full CD pipeline view—but not all the stages of the pipeline are implemented.

Fig. 3 shows an example. The first row in the figure represents the CD pipeline. Each gray box in the row represents a stage in the pipeline. The pipeline runs from left to right. The transition from one stage to another is automatic except for the transition to user acceptance and the transition to production. Each code change triggers an execution of the pipeline. Each execution of the pipeline is represented as a row under the first row. (For brevity, I listed only three executions.) Each box in the row indicates the status of a stage. The initial color is blue, which indicates the stage has not been executed. When the code change passes a stage of the pipeline, the box that represents the stage turns green. When the code change fails to pass the stage, the box that represents the stage turns red.

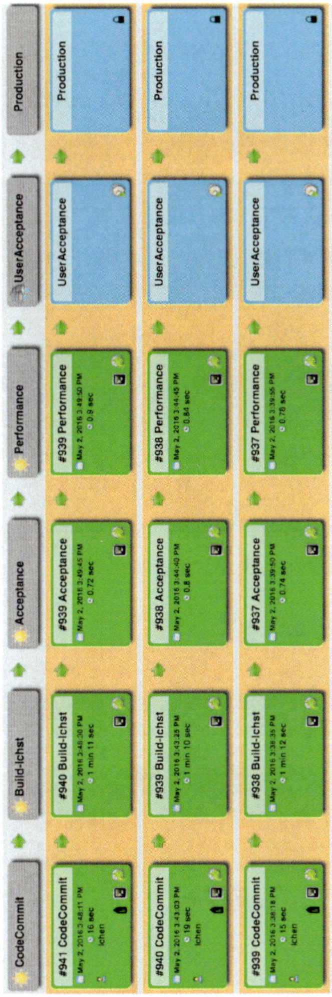

Figure 3: An example visual CD pipeline skeleton.

In this example, the team does not have automated acceptance tests and performance tests. The application's deployment cannot be fully automated either due to a problem in their architecture. Making improvements in the above areas can take a long time. We give the team a visual pipeline skeleton that has the same look and feel as the final pipeline, but the acceptance and performance stages do not do anything (you can see it takes less than one second to complete). Similarly, the user acceptance and production stages have not yet been implemented. Although the pipeline is not complete, it helps to maintain the team's CD adoption momentum. Its visual nature makes the abstract concept of CD tangible. The pipeline sets a clear and visible vision and roadmap so that all the stakeholders can track the progress of ongoing improvements.

The pipeline is a thing that developers see every day, and it acts as a reminder. Every time the developers see the pipeline view, the gaps in their CD adoption are clear. For example, they can easily see that the acceptance and performance stages are empty. We have often seen developers start discussions about how they could fill the gaps by writing automated acceptance tests and performance tests.

Because the pipeline is already in place, the teams can begin enjoying the benefits of any improvements they make toward CD immediately. For example, whenever they write some automated tests and add them, the pipeline will begin to automatically execute the tests after each code commit. The immediate feedback from these benefits will further encourage the developers to write more automated tests, helping to maintain the CD adoption momentum.

Expert drop (S6)

Another strategy to address tough applications is expert drop. With expert drop, we assign an expert in CD adoption to join the team facing difficulties in migrating their tough application to CD. This expert will help to drive the changes from within the team in areas such as team practices, team culture, and applications architecture to help ease the migration.

The person in this expert role is a senior member of a team, possesses deep knowledge about CD, and has been through the full CD migration process. As mentioned earlier, by the time we started to onboard the tough applications, we had already successfully moved several important applications to CD; therefore, the experts were selected from the teams who had already gone through the migration process.

The use of expert drop is important for some applications; trying to implement the required changes from outside a team may not always work. For example, to accelerate the adoption of CD by application development teams, the top management team started an initiative. A group of experts was formed who would visit the teams and help them to make the required improvements. When a team was selected, training sessions were organized to teach the team the "correct" way of working.

Every day, there was a stand-up meeting with all the experts and the team members to push the improvements forward.

This group of experts had been helping one team with a tough application make changes for several months. However, the team members were resistant. Little progress had been achieved. When we talked to the members of the team, we found that they could not understand why they needed to change; they were happy with their old way of working. They also commented that the external experts did not understand their problems.

That led us to change the strategy. Rather than pushing the team to change using a group of external experts, we assigned an expert to join the team as a senior member. Then, the expert started to drive the changes from within the team.

Because he was a part of the team on a day-to-day basis, he discussed, communicated, and interacted with the members of the team. In the process, he obtained deeper insights into the problems the team faced, allowing him to explain to the team members why they needed to change their existing work processes in a way that the team members understood. He was also able to explain various CD practices in the context of a specific problem faced by the team.

The results were positive. The team members began to embrace CD. They started to exhibit interest in moving some CD adoption tasks forward. Gradually, the momentum to make changes was built. The release frequency increased, and the requirements and testing practices became more aligned with the requirements of CD.

Why did the team members collaborate with the senior member when she was a member of the team but were reluctant to collaborate with her when she was an external expert? Team culture is one factor. Overall our organization promotes collaboration, which is one of the company's five core values. However, in a large organization, in a real industrial setting, teams with culture problems can exist. Although the number of such teams is small, when we aim at implementing CD across the whole organization, we must deal with them.

Team culture is just one aspect of the difficulties. In some cases, due to the complexities and diversities of the situations, the experts, looking from outside the team, could not correctly and deeply understand the problems. Thus, the advice they gave may have sounded good but were not actually feasible to implement.

The strategy of expert drop to drive changes from within the team can be useful for migrating tough applications, especially when imposing the migration from outside the team does not work well. Using this strategy, we have successfully migrated some very tough applications to CD, some of which had already failed an initial migration attempt using an external expert group.

CHALLENGES FOR RESEARCH

What we know now about CD adoption is only the tip of the iceberg. There are many more specific challenges to address. Below, I identify some important challenges that need researchers' help. These challenges are important because they are likely to be faced by many practitioners trying to adopt CD. Any improvements (research breakthroughs) in these areas could result in significant benefits to practitioners.

Organizational challenges

I mentioned organizational challenges in my earlier paper (Chen, 2015a). However, apart from the points discussed in Chen (2015a), I recently noticed another area in which evidence-based results are important. Organizational structures can be a significant barrier when implementing CD (Chen, 2015a). We must change the existing organizational structure to create successful implementations. However, there is no target organizational structure that is evidence-based and validated. Therefore, new organizational structures are often based solely on opinion. As an example, our previous CTO re-structured the organization in one way. A new CTO took the job recently and had a different opinion. He re-structured the organization in yet another way—in fact, he reverted some of the changes the previous CTO had implemented.

Although there is a large corpus of studies on organizational structures (Organizational Structure, 2016), we still need evidence-based results regarding the connections between CD and organizational structures. For example, what are the possible structures that suit CD? What are the advantages and disadvantages of each? Which structures are most suitable for which situations? What are the organizational design guidelines? We need this information to make evidence-based decisions.

Process

I also mentioned process challenges in my earlier work (Chen, 2015a). However, to the best of my knowledge, these challenges have not received sufficient attention from researchers. Therefore, here I provide additional rationale on why I think processes are an important topic of study.

Most large organizations have many processes that must be followed to release a unit of software to production. For example, in our organization, we needed to create a change ticket, place the change request on the agenda of the next Change Advisory Board (CAB) meeting (Rob, 2007), present the change at the meeting, receive CAB approval, confirm the deployment window, and so forth.

This is particularly problematic at very large enterprises with long histories. For example, I have met practitioners from these large enterprises who report that they needed to provide more than one hundred different documents and follow numerous processes to release any code change to production. Worse, these processes could take several weeks or even months. Such lengthy processes prevent companies from adopting CD. We must change the existing processes.

However, changing these processes is very difficult because they were originally put in place to ensure quality and security. Many of them were introduced to prevent previous incidents from happening again. In fact, many organizations employ people whose sole job is to ensure that everyone follows the official processes.

Nevertheless, to achieve CD, we must change the processes. Although some practitioners have reported anecdotal stories, no systematic approach or strategies are available for other practitioners to follow. This is a gap that remains to be filled.

Tools

Implementing CD requires a new toolchain because, as reported by several studies, existing tools are inhibitory in achieving CD (Olsson et al., 2012; Rissanen and Münch, 2015; Marschall, 2007; Zhu et al., 2015; Debbiche et al., 2014; Volker et al., 2013). We call this new toolchain the "CD Platform".

This platform is a fairly complex software system. It must integrate with a large number of other tools, including version control systems, artifact repositories, release management systems, application and machine configuration management systems, deployment tools, identity management tools, and so on. It also needs to have the following quality:

- Reliable, because all the developers will rely on it to carry out their work.
- Secure, because production systems will be in danger if it is subverted (Bass et al., 2015).
- Scalable, to support the continually increasing number of applications in an organization.
- Easy to use, so that a team can implement a new deployment pipeline in a self-service manner; otherwise the CD team will likely be unable to support a large number of application development teams.

However, compared to research efforts spent on tools for testing, requirements, architecture, etc., tools for CD have not attracted enough attention from researchers. As a practitioner, if I am going to buy a tool for CD, there is no evaluation framework that I can use to evaluate and select from the available tools and vendors. If I am going to build the tool in-house, there is no reference architecture available. When I design the pipeline stages, there are no guidelines or best practices to follow either. Therefore, I would like to draw more researchers' attention to this area.

Infrastructure

Implementing CD requires infrastructure support. By infrastructure, I mean computing power (e.g., CPU), storage, network, memory, etc., all of which are typically managed by the infrastructure team. Infrastructure has been mentioned as a challenge by some studies. For example, Claps et al. (2015) reported that the need for adding additional computing power, network bandwidth, and memory to CI servers is a challenge. They consider adding more resources as a strategy (Claps et al., 2015).

However, according to our experience, the preceding problems do not provide a full picture of the infrastructure challenges. The way the infrastructure is managed is likely to be an even greater challenge. Traditionally most infrastructure management is largely manual. When a team needs a server, they make a request to the infrastructure team. The infrastructure team creates a Virtual Machine (VM) for them. The system engineering team installs and configures the appropriate operating system on that VM. The network team adds the VM onto the network. At this stage, the new VM does not have access to other machines on the network because, by default, communications are blocked by the firewall for security reasons. The team then creates a change request to update the firewall rules. The security team reviews and approves the change request. The network team executes the change request. The VM is then handed over to the application operations team, which installs and configures the application on the VM. The system engineering team configures the load balancer to make the application on the VM available to external customers. Because many teams are involved in this process, the whole process can take several weeks or a month. This method of managing infrastructure is a hindrance to companies attempting to achieve CD.

To support CD, ideally we need the capability to automatically provision the required infrastructure. In other words, we need to change to Infrastructure as a Service (IaaS) (Mell and Grance, 2011).

However, making a company's entire infrastructure available as a service is a big transformation for any organization, and the transformation itself can take a very long time (e.g., a couple of years). For example, in our company, due to the large cash investment required, the decision was made very carefully. Selecting the technologies, investigating the feasibility, and evaluating different vendors took a long time. When a decision was finally made, implementing it also required a long time. Apart from the technical implementation, the workflow of infrastructure management also needed to be updated.

According to informal conversations with attendees at the DevOps Showcase Dublin conference (http://conferences.unicom.co.uk/fundamentals-of-devops/), the Delivery of Things World conference (http://deliveryofthingsworld.com/en/), and the QCon London conference (https://qconlondon.com/), I found that for many companies, infrastructure management methods are not ideal for CD. Many organizations have begun to discuss or even implement IaaS, but not all of them have actually turned their entire infrastructure into a service that can be managed and consumed by APIs. Even moving to a public cloud such as AWS may not be as straightforward as one might wish. For example, as reported by an attendee of the DevOps Showcase Dublin conference, an Irish ecommerce company had to move their services from AWS back to their internal data center because the application required significant amounts of I/O Operations (IOPs). The price for IOPs on AWS was too expensive; running the services in their own data center proved to be a much cheaper option.

Consequently, implementing CD is a problem when the desired infrastructure support is not ready. This is an important topic to address because many companies that would like to adopt CD are in similar situations.

However, there are no studied approaches, strategies, or guidelines in the literature to address this problem. Part of our future work is to derive and formulate some guidelines and strategies from the experiences of companies that have managed to overcome their infrastructure problems.

Legacy systems

According to our experience, there are probably more legacy systems than green field projects in the industry (Bennett, 1995). These legacy systems' architectures are usually not amenable to CD (Chen, 2015b). Moreover, the teams working on these legacy systems usually have "legacy" culture, practices, and mindsets that would need to be changed to migrate the applications to CD. Unfortunately, making these changes is very difficult. Considering the large number of such applications in the industry, studying the strategies and analyzing the approaches for moving legacy systems to CD is very important because any improvements in this area could have significant practical impacts. However, little, if any, work has been reported that specifically seeks to understand and address this important type of software (Rodríguez et al., 2016).

Architecting for continuous delivery

We found that application architecture can be a key barrier for CD adoption. However, discussions of CD have primarily focused on build, test, and deployment automation; insufficient attention has been paid to application architecture. More research is needed to provide deeper understanding and guidance to help practitioners (re-) architect their applications for CD (Chen, 2015b).

Continuous testing of non-functional requirements

While unit tests and acceptance tests have been extensively discussed and widely practiced in CD, testing non-functional requirements has received considerably less attention. However, unit tests and acceptance tests are mainly intended to ensure functional requirements.

Performing only functional testing might be acceptable for some applications that are insensitive and tolerant of non-functional requirements. However, there are many applications that are strict and sensitive to non-functional requirements. For example, we have many applications that have strict performance requirements measured in milliseconds.

There are extra challenges in testing non-functional requirements. For example, with functional requirements we can easily determine whether a test passes or fails. However, with some non-functional requirements, defining the criteria for passing or failing a test is less straightforward. For example, with a performance test, the results can vary from one test to another due to unavoidable variations in the testing environments, such as network conditions.

Testing non-functional requirements may require a larger scale testing environment than functional testing. For example, scalability testing often needs a larger scale environment than acceptance testing.

Automating tests for non-functional requirements can be quite different from writing tests for functional requirements. For example, when testing reliability, the tests need to create different failure scenarios such as killing some application nodes or creating network partitions between nodes, etc.

Thus, testing of non-functional requirements in the context of CD is an important challenge. I believe this is another interesting area for researchers to explore.

Test execution optimization

When practicing CD, software releases occur much more frequently. Before CD, an application might typically have fewer than six releases per year. Under CD, an application releases at least once a week and sometimes multiple times per day.

To ensure the quality of each release, tests are executed more frequently. Unit tests are executed for every code commit. Acceptance tests are executed for each code commit that passes the unit tests.

This significantly increased frequency of test execution makes test optimization especially important and useful. If even a small saving can be achieved for each test execution, when multiplied by the large number of executions, the total savings quickly become significant.

Moreover, the potential saving we could get from each test execution is significantly bigger in a CD context than in a traditional context. Before CD, when we released every six months, the number of changes that each release included was large. In each test execution, to cover all the changes, it is likely that many test cases needed to be executed. In contrast, under CD, the number of changes in each release is small. To cover this significantly smaller number of changes, it is likely that only a much smaller set of tests needs to be executed. However, we still execute the tests in the old way—in other words, we still execute all the test cases even when only one line of code has been changed. Thus, there is plenty of room for test optimization under CD.

Researchers have proposed some test selection techniques and test prioritization techniques, but these proposals have not been widely adopted in industry (Singh et al., 2012; Engström et al., 2010; Gligoric, 2015). We need to identify the gaps and develop new approaches or improve the existing approaches to fill those gaps.

In addition, existing test selection or test prioritization techniques mainly focus on optimizing an individual test execution. The high frequency of test executions in CD could make optimizing across multiple adjacent test executions a viable option to explore. For example, by knowing that we normally run tests ten times a day, is it possible to reduce a single test case's repetition across all these ten test executions?

Finally, as the popularity of CD increases, test optimization techniques also become more important in practice. I would like to draw more researchers' attention to test optimization as well.

RELATED WORK

In this section, I highlight the differences of this paper in relation to existing literature and to my previous work.

Differences from existing strategies

Many works have reported challenges and solutions in adopting CD (Leppanen et al., 2015; Claps et al., 2015; Olsson et al., 2012; Karvonen et al., 2015; Rissanen and Münch, 2015; Marschall, 2007; Zhu et al., 2015; Debbiche et al., 2014; Puneet, 2011; Noureddine and Foutse, 2014; Laukkanen et al., 2015; Rogers, 2004; Sekitoleko et al., 2014; Krusche and Alperowitz, 2014; Adams et al., 2015; Souza et al., 2015; Feitelson et al., 2013; Rahman et al., 2015; Neely and Stolt, 2013), including a very recent systematic literature review (Laukkanen et al., 2017). However, none of the existing literature on CD has included the strategies I reported in this paper.

Nevertheless, I believe these strategies are useful because we found that putting some of the existing strategies into action is challenging in its own right. For example, Claps et al. (2015) reported that it is important to "ensure that top management implements a strategy to push the need to implement the CD process" as a strategy to adopt CD. However, putting this strategy into action is a challenge because the CD implementation team is typically not in a position to order top management to implement any strategy to push the CD implementation process forward. Instead, the real challenge for us was how to acquire their buy-in and sustained support, so that we could convince them to carry out that action.

As another example, Claps et al. (2015) also reported providing more resources to a product's CI servers as a strategy to adopt CD. However, acquiring more resources was a true challenge for us. Without buy-in by the related stakeholders and their sustained support, we were unable to obtain the required resources.

As a third example, several papers have reported moving to more flexible and modular designs as a strategy to adopt CD (Claps et al., 2015; Olsson et al., 2012; Debbiche et al., 2014; Noureddine and Foutse, 2014; Sekitoleko et al., 2014; Laukkanen et al., 2017). However, moving a legacy system to a flexible and modular design is a major challenge and may require significant effort over a long period of time. The CD implementation team does not own the applications and cannot make changes to them. Instead, our challenge was to convince the responsible teams to make the changes and find ways to maintain their momentum until those changes were complete. The strategies reported in this paper aim at addressing these challenges so that we can make the required changes happen.

Differences to change management literature

Adopting CD is a change process. Change management has been studied and a large corpus of change management literature exists (Todnem By, 2005; Change Management, 2016).

For some of the strategies I reported, a generalized form can be found in the general change management literature. For example, we could say a more generalized form for S3 (continuous delivery of continuous delivery) is the strategy of making changes incrementally rather than all at once. Thus, the strategy may seem too general.

However, I have described these strategies in the specific context of CD adoption. I also provided experience-based advice specific to CD adoption. For example, in S2 (establishing a dedicated team with multi-disciplinary members), I provided experience-based advice on what kinds of disciplines should be included in the multi-disciplinary team and why we need each of those disciplines. For another example, in S3 (continuous delivery of continuous delivery), we provided experience-based heuristics on how to decompose the whole CD adoption effort into smaller pieces in such a way that we can complete each piece in a short time and gain the benefits from that piece immediately. As a third example, in S4 (starting with easy but important applications), I provided heuristics on what applications are easy to migrate.

Some strategies I reported are specific to CD adoption. For example, S5 (visual CD pipeline skeleton), for which we need a CD pipeline to use it.

Thus, although literature on change management exists, it does not cover the specifics of introducing CD to an organization. I believe other practitioners will find the CD-specific details I provided in this paper useful.

Differences from existing reported challenges

The strategies reported in this paper aim at addressing the challenges of obtaining buy-in from a wide range of stakeholders whose support is required when implementing CD, gaining sustained support to continue the often lengthy and tough journey toward CD, and maintaining the momentum of application teams when their CD migrations require huge efforts over long periods. These challenges are not reported in the recent Systematic Literature Review (SLR) (Laukkanen et al., 2017), which systematically reviewed the existing literature to identity problems and solutions in adopting CD.

I also outlined a few further research challenges that are not reported in the above-mentioned SLR, such as process challenges, CD platform development challenges, and infrastructure management challenges. For those challenges that were mentioned before, in this paper I provided extra new descriptions that enrich our understanding of the challenges.

Differences from my previous work

My previous work (Chen, 2015a) did not include any of the strategies reported in this paper. In terms of challenges, four of the further challenges for research were not included in my previous paper. For those challenges that were mentioned in (Chen, 2015a), I provided new and additional descriptions in this paper.

In addition, this work was based on four year's longitudinal first-hand experiences in implementing CD across an entire large organization. The first-hand experiences allowed us to feel the real pain and develop a richer and deeper understanding. Over a long time helped us to encounter challenges that are revealed only at the far end of the journey to CD (e.g., processes were not seen as challenges initially). Implementing CD across the entire large organization helped us to see challenges relating to team and application varieties (e.g., there are very tough applications and teams). Our context is different from that of other papers and has helped us to identify some of the challenges that have not been reported before.

SUMMARY

Continuous Delivery (CD) has become a popular software development approach. Motivated by the huge benefits reported (Chen, 2015a; Leppanen et al., 2015), many practitioners want to implement CD in their organizations (Perforce Software Inc., 2015). However, implementing CD can be very challenging (Chen, 2015a; Leppanen et al., 2015; Claps et al., 2015).

To help overcome CD adoption challenges, this paper reported six strategies derived from our four-year journey to implement CD at a multi-billion-euro company. These strategies are summarized in Table 2.

Table 2: Summary of strategies to overcome adoption challenges.

Strategy	Description
Selling CD as a painkiller	Identify each stakeholder's pain points that CD can solve, and sell CD as a painkiller to that stakeholder. This strategy helps to achieve buy-in from the wide range of stakeholders that a CD implementation requires.
Dedicated team with multi-disciplinary members	Without a dedicated team, it can be hard to progress because employees are often assigned to work on other value streams. A multi-disciplinary team not only provides the wide range of skills required for CD implementation but also smooths the communication with related teams.
Continuous delivery of continuous delivery	Organize the implementation of CD in a way that delivers value to the company as early as possible, onboarding more projects gradually, in small increments and eventually rolling out CD across the whole organization. This strategy helps justify the investment required by making concrete benefits visible along the way. Visible benefits, in turn, help to achieve the sustained company support and investment required to survive the long and tough journey to CD.
Starting with easy but important applications	When selecting the first few applications to migrate to CD, choose the ones that are easy to migrate but that are important to the business. Being easy to migrate helps to demonstrate the benefits of CD quickly, which can prevent the implementation initiative from being killed. Being important to the business helps to secure the required resources, demonstrates clear and unarguable value, and raises the visibility of CD in the organization.

Visual CD pipeline skeleton	Give a team a visual CD pipeline skeleton that has the full CD pipeline view but with empty stages for those they cannot implement yet. This helps to build up a CD mindset and maintain the momentum for CD adoption. The pipeline skeleton is especially useful when the team's migration to CD requires a large effort and mindset changes over a long period of time.
Expert drop	Assign a CD expert to join tough projects as a senior member of the development team. Having the expert on the team helps to build the motivation and momentum to move to CD from inside the team. It also helps to maintain momentum when the migration requires a large effort and a long period of time.

I hope these strategies can help fellow practitioners to overcome similar challenges and reap the huge benefits of CD (Chen, 2015a; Leppanen et al., 2015). The information also contributes toward building a body of knowledge concerning CD adoption.

For many practitioners, CD is a clear goal, but achieving that goal is difficult. What we currently know about CD adoption is just the tip of the iceberg. There are many more specific challenges to address, such as the challenges discussed in this paper regarding organizational structure, processes, tools, infrastructure, legacy systems, architecting for CD, continuous testing of non-functional requirements, and test execution optimization. I invite more researchers to investigate these increasingly important topics.

ACKNOWLEDGMENTS

I thank Dave Farley of Continuous Delivery Ltd for reviewing these strategies and my colleagues at Paddy Power for their support. I also thank the reviewers for their thoughtful comments and suggestions to improve this paper. This paper is based on my experience at Paddy Power. It represents only my own views and does not necessarily reflect those of the employer.

REFERENCES

1. Adams, B , Bellomo, S. , Bird, C. , Marshall-Keim, T. , Khomh, F. , Moir, K. , 2015. The practice and future of release engineering: a roundtable with three release en- gineers. IEEE Software 32, 42–49 .
2. Anderson, D.J. , 2010. Kanban: Successful Evolutionary Change for Your Technology Business. Blue Hole Press .
3. Bass, L. , Holz, R. , Rimba, P. , Tran, A.B. , Zhu, L. , 2015. Securing a deployment pipeline. Presented at the 2015 IEEE/ACM 3rd International Workshop on Release Engi- neering (RELENG) .
4. Beck, K. , 2002. Test Driven Development: By Example. Addison-Wesley Professional, Boston, MA .
5. Beck, K., Beedle, M., Van Bennekum, A., Cockburn, A., Cunningham, W., Fowler, M., et al., 2001. Manifesto for agile software development.
6. Bennett, K. , 1995. Legacy systems: coping with success. IEEE Software 12, 19–23 .
7. Change Management (2016). Available: https://en.wikipedia.org/wiki/Change _ management (Accessed 30.10.16).
8. Chen, L. , 2015a. Continuous delivery: huge benefits, but challenges too. IEEE Soft- ware 32, 50–54 .
9. Chen, L. , 2015b. Towards architecting for continuous delivery. In: 2015 12th Working IEEE/IFIP Conference on Software Architecture (WICSA), pp. 131–134 .
10. Claps, G.G. , Berntsson Svensson, R. , Aurum, A. , 2015. On the journey to continuous deployment: technical and social challenges along the way. Inform. Software Tech. 57, 21–31 .
11. Debbiche, A. , Dienér, M. , Berntsson Svensson, R. , 2014. Challenges when adopt- ing continuous integration: a case study. In: Jedlitschka, A., Kuvaja, P., Kuhrmann, M., Männistö, T., Münch, J., Raatikainen, M. (Eds.). In: Product-Fo- cused Software Process Improvement, vol. 8892. Springer International Publish- ing, pp. 17–32 .
12. Donovan, A. , Kernighan, B.W. , 2015. The Go Programming Language. Addison-Wes- ley .

13. Engström, E. , Runeson, P. , Skoglund, M. , 2010. A systematic review on regression test selection techniques. Inform. Software Tech. 52, 14–30 .
14. Feitelson, D.G. , Frachtenberg, E. , Beck, K.L. , 2013. Development and deployment at Facebook. IEEE Internet Comput. 17, 8–17 .
15. Gligoric, M.Z. , 2015. Regression Test Selection: Theory and Practice. University of Illinois at Urbana-Champaign .
16. Humble, J. , Farley, D. , 2010. Continuous Delivery: Reliable Software Releases through Build, Test, and Deployment Automation. Addison-Wesley, New York .
17. Karvonen, T. , Lwakatare, L.E. , Sauvola, T. , Bosch, J. , Olsson, H.H. , Kuvaja, P. , et al. , 2015. Hitting the target: practices for moving toward innovation experiment systems. In: Fernandes, M.J., Machado, J.R., Wnuk, K. (Eds.), Software Business: 6th international conference, ICSOB 2015, Braga, Portugal, June 10-12, 2015, pro- ceedings. Cham: Springer International Publishing, pp. 117–131 .
18. Krusche, S. , Alperowitz, L. , 2014. Introduction of continuous delivery in multi-cus- tomer project courses. In: Presented at the Companion Proceedings of the 36th International Conference on Software Engineering (ICSE Companion 2014). Hy- derabad, India .
19. Laukkanen, E. , Itkonen, J. , Lassenius, C. , 2017. Problems, causes and solutions when adopting continuous delivery—a systematic literature review. Inf. Software Tech- nol. 82, 55–79 2// .
20. Laukkanen, E. , Paasivaara, M. , Arvonen, T. , 2015. Stakeholder perceptions of the adoption of continuous integration – a case study. In: 2015 Agile Conference (AGILE), pp. 11–20 .
21. Leppanen, M. , Makinen, S. , Pagels, M. , Eloranta, V.-P. , Itkonen, J. , Mantyla, M.V. , et al. , 2015. The highways and country roads to continuous deployment. IEEE Software 32, 64–72 .
22. Marschall, M. , 2007. Transforming a six month release cycle to continuous flow. In: 2007 Agile Conference (AGILE), pp. 395–400 .
23. Mell, P. , Grance, T. , 2011. The NIST definition of cloud computing. Computer Security Division, Information Technology Laboratory.

National Institute of Standards and Technology, Gathersburg, MD .

24. Neely, S. , Stolt, S. , 2013. Continuous delivery? Easy! Just change everything (Well, maybe it is not that easy). In: 2013 Agile Conference (AGILE), pp. 121–128 .
25. Noureddine, K. , Foutse, K. , 2014. Factors impacting rapid releases: an industrial case study. In: Presented at the Proceedings of the 8th ACM/IEEE International Symposium on Empirical Software Engineering and Measurement (ESEM '14).
26. Torino, Italy .
27. O'Dell, C. , Skelton, M. , 2016. Continuous Delivery with Windows and .NET. O'Reilly .
28. Odersky, M. , Spoon, L. , Venners, B. , 2016. Programming in Scala: A Comprehensive Step-by-Step Guide, third ed. Artima Press .
29. Olsson, H.H. , Alahyari, H. , Bosch, J. , 2012. Climbing the "Stairway to Heaven" – a mulitiple-case study exploring barriers in the transition from agile development towards continuous deployment of software. In: 2012 38th EUROMICRO Confer- ence on Software Engineering and Advanced Applications (SEAA), pp. 392–399 .
30. Organizational Structure. Available: https://en.wikipedia.org/wiki/ Organizational _ structure . (Accessed 30.10.16).
31. Perforce Software Inc, 2015. Continuous Delivery: The New Normal for Software Development Available: http://info.perforce. com/continuous- delivery- survey- report.html .
32. Puneet, A. , 2011. Continuous SCRUM: agile management of SAAS products. In: Pre- sented at the Proceedings of the 4th India Software Engineering Conference (ISEC '11). Thiruvananthapuram, Kerala, India .
33. Rahman, A .A .U. , Helms, E. , Williams, L. , Parnin, C. , 2015. Synthesizing continuous deployment practices used in software development. In: 2015 Agile Conference (AGILE), pp. 1–10 .
34. Rissanen, O. , Münch, J. , 2015. Transitioning towards continuous delivery in the B2B domain: a case study. In: Lassenius, C., Dingsøyr, T., Paasivaara, M. (Eds.), Ag- ile Processes in Software Engineering and Extreme Programming: 16th Inter- national Conference, XP 2015, Helsinki, Finland, May 25-29, 2015, Proceedings. Cham:

Springer International Publishing, pp. 154–165 .

35. Rob, A. , 2007. Effective IT Service Management: To ITIL and Beyond!. Springer-Ver- lag, New York, Inc .
36. Rodríguez, P. , Haghighatkhah, A. , Lwakatare, L.E. , Teppola, S. , Suomalainen, T. , Es- keli, J. , et al. , 2016. Continuous deployment of software intensive products and services: a systematic mapping study. J. Syst. Software (January) .
37. Rogers, R.O. , 2004. Scaling Continuous Integration. In: Eckstein, J., Baumeis- ter, H. (Eds.), Extreme Programming and Agile Processes in Software Engineer- ing: 5th International Conference, XP 2004, Garmisch-Partenkirchen, Germany, June 6-10, 2004. Proceedings. Springer Berlin Heidelberg, Berlin, Heidelberg, pp. 68–76 .
38. Savor, T. , Douglas, M. , Gentili, M. , Williams, L. , Beck, K. , Stumm, M. , 2016. Continuous deployment at Facebook and OANDA. In: Presented at the Proceedings of the 38th International Conference on Software Engineering Companion (ICSE '16).
39. Austin, Texas .
40. Sekitoleko, N. , Evbota, F. , Knauss, E. , Sandberg, A. , Chaudron, M. , Olsson, H.H. , 2014. Technical dependency challenges in large-scale agile software develop- ment. In: Cantone, G., Marchesi, M. (Eds.), Agile Processes in Software Engineer- ing and Extreme Programming: 15th International Conference, XP 2014, Rome, Italy, May 26-30, 2014. Proceedings. Cham: Springer International Publishing, pp. 46–61 .
41. Singh, Y. , Kaur, A. , Suri, B. , Singhal, S. , 2012. Systematic literature review on regres- sion test prioritization techniques. Informatica 36 .
42. Souza, R. , Chavez, C. , Bittencourt, R.A. , 2015. Rapid releases and patch backouts: a software analytics approach. IEEE Software 32, 89–96 .
43. Todnem By, R. , 2005. Organisational change management: a critical review. J. Change Manage. 5, 369–380 .
44. Volker, G. , Christoph, H. , Christian, J. , 2013. Security of public continuous integration services. In: Presented at the Proceedings of the 9th International Symposium on Open Collaboration (WikiSym '13). Hong Kong, China .

45. Wynne, M. , Hellesoy, A. , 2012. The Cucumber Book: Behaviour-Driven Development For Testers and Developers. Pragmatic Bookshelf, Dallas, TX .

46. Zhu, L. , Xu, D. , Tran, A.B. , Xu, X. , Bass, L. , Weber, I. , et al. , 2015. Achieving reliable high-frequency releases in cloud environments. IEEE Software 32, 73–80 .

CHAPTER 11

HOW DO SOFTWARE DEVELOPMENT TEAMS MANAGE TECHNICAL DEBT? – AN EMPIRICAL STUDY

Jesse Yli-Huumo[a], Andrey Maglyas [a], Kari Smolander [b]

[a] Lappeenranta University of Technology, School of Business and Management, Department of Innovation and Software, PO Box 20, Skinnarilankatu 34, Lappeenranta FI-53851, Finland

[b] Aalto University, School of Science, Department of Computer Science, P.O.Box 15400, FI-00076 Aalto, Finland

ABSTRACT

Technical debt (TD) is a metaphor for taking shortcuts or workarounds in technical decisions to gain short-term benefit in time-to-market and earlier software release. In this study, one large software development organization is investigated to gather empirical evidence related to the concept of technical debt management (TDM). We used the exploratory case study method to collect and analyze empirical data in the case organization by interviewing a total of 25 persons in eight software development teams. We were able to identify teams where the current strategy for TDM was only to fix TD when necessary, when it started to cause too much trouble for development. We also identified teams where the management had a systematic strategy to identify, measure and monitor TD during the development process. It seems that TDM can be associated with a similar maturity concept as software development in general. Development teams may raise their maturity by increasing their awareness and applying more advanced processes, techniques and tools in TDM. TDM is an essential part of sustainable software development, and companies have to find right approaches to deal with TD to produce healthy software that can be developed and maintained in the future.

Keywords Technical debt, Technical debt management, Exploratory case study

INTRODUCTION

Technical debt (TD) is a metaphor used to describe a situation in software development, where a shortcut or workaround is used in a technical decision (Kruchten et al., 2012b). TD has also similarities to three aspects of financial debt: repayment, interest, and in some cases high cost (Allman, 2012). In software development, a shortcut or workaround can give the company a benefit in the short term with quicker release to the customer and an advantage in time-to-market over the competition (Kruchten et al., 2012a; Yli-Huumo et al., 2015a). However, if these shortcuts and workarounds are not repaid, TD can accumulate and hurt the overall quality of the software and the productivity of the development team in the long term (Zazworka et al., 2011b). Creating temporary solutions to the code base increases complexity, which makes further development hard and time-consuming (Yli-Huumo et al., 2015a; Yli-Huumo et al., 2014). A simple solution for the problem would be to repay the known TD before issues start to show. However, the highly competitive software market forces companies to work in tight schedules and deadlines to release software to customers in faster cycles. This creates constant pressure for the development teams to deliver working features to customers within the given deadlines. In addition, perfection as an objective is also a risk, because it may cause delays and that way frustration to the customers, who may then select other commercial alternatives. Therefore, it is important to identify and develop processes for companies to live with TD and to know how, what and when the TD should be repaid. Technical debt management (TDM) consists of activities, processes, techniques, and tools that can be used to identify, measure, prevent, and reduce TD in a software product.

TD and TDM receive attention currently both in the academia and the industry (Li et al., 2015a). Researchers and practitioners are becoming more interested in the concept of TD and the reasons why it should be an essential part of decision-making in software development (Falessi et al., 2014). The current literature has identified and developed some tools and practices to conduct TDM. However, according to a recent mapping study, the problem is the lack of empirical evidence about TDM in a real-life software development environment (Li et al., 2015a). It is important to gather evidence about TD and TDM in real-life software development situations to understand how TDM is currently perceived by real software development teams, and to use that knowledge to improve the existing processes and tools.

In order to understand TDM in a real-life software development environment, we studied eight software development teams in a large organization that is a provider of multiple software solutions. For data collection and analysis, we used the eight TDM activities identified by Li et al. (2015a) in semi-structured interviews to gather empirical data about TDM in the selected software development teams. We used the exploratory case study method (Robson, 2002) to answer the following main research question:

RQ: "How do software development teams manage technical debt?"

Since the main research question can be considered quite a wide topic, including several other topics, we decided to create a set of sub questions to tackle specific topics of our interest.

RQ1.1: What TDM activities are used in the studied development teams?

Technical debt management can be separated into the following activities: identification, measurement, prioritization, prevention, monitoring, repayment, representation/documentation, and communication (Li et al., 2015a). However, it is not certain what activities are actually used and taken into consideration in real-life software development. Therefore, it is important to study and understand which TDM activities are currently applied/used and which are not. The results obtained from the studied development teams could reveal which activities will need more research in the future.

RQ1.2: What methods, models, practices or tools do the studied development teams use for each TDM activity?

There are a number of possible methods, models, practices or tools for every TDM activity (Li et al., 2015a). They have been developed and suggested in the literature, but they lack empirical evidence of their usability and functionality (ibid.). Therefore, it is essential to gather empirical evidence from real-life software development to understand what approaches different software development teams use for each TDM activity. Collecting such evidence could help to evaluate which TDM approaches should be categorized to each TDM activity.

RQ1.3: Are there any maturity differences in adopting TDM activities between development teams?

Every software development team is different, working with different products in different environments, and using different methods, models, practices, and tools in their unique way. It is highly possible that software development teams in general have different activities and approaches as regards TDM. Some software development teams may use more time on TDM, while some development teams may not pay much attention to it (Power, 2013). Therefore, it is important to understand if it is possible to distinguish between different maturities of TDM, similarly as in the capability maturity model (CMM)(Paulk et al., 1993). The results of this study can be used to develop a similar maturity model for TDM, which researchers and practitioners could use to conduct more research, or to improve companies' internal and external practices.

RQ1.4: What are the biggest challenges in TDM?

Software process improvement includes the challenge of adopting new practices and tools to development teams. Understanding this challenge in relation to TDM is beneficial for software development teams and researchers.

The rest of the paper is organized as follows: Section 2 introduces the theoretical background of TD and TDM in software development, Section 3 describes the research methodology used in this study, and Section 4 presents the results received from the empirical analysis of the studied software development teams. In Section 5present the developed framework. In Section 6 we discuss the results and implications to future research. Section 7 concludes the paper.

BACKGROUND

Technical debt

The metaphor technical debt (TD) has been introduced by Ward Cunningham (Cunningham, 1992). He describes the metaphor as 'Shipping first time code is like going into debt. A little debt speeds development so long as it is paid back promptly with a rewrite. Objects make the cost of this transaction tolerable. The danger occurs when the debt is not repaid. Every minute spent on not-quite-right code counts as interest on that debt."(op.cit., p. 29-30). Even though the metaphor was first introduced over twenty years ago, a recent mapping study shows that it has received the attention of researchers and practitioners only in the past few years (Li et al., 2015a).

The TD metaphor was first associated with compromises on the code level of software (Cunningham, 1992). In addition, terms like *code smells* (Fowler et al., 1999) have described situations where poor technical choices in software development have caused problems in code quality and architectural soundness. However, the TD metaphor has been rapidly expanded after the initial concept on the code level, and it has been associated with other stages of the software development lifecycle as well (Tom et al., 2013; Alves et al., 2014). The current literature identifies such terms as requirements (Brown et al., 2010), design (Zazworka et al., 2011b; Zazworka et al., 2011a), architectural (Nord et al., 2012), test (Brown et al., 2010), process (Lim et al., 2012), documentation (Kruchten et al., 2012a), and people debt (Kruchten et al., 2012b) to demonstrate the same effect of shortcuts or workarounds happening in the other stages of the software development lifecycle.

Shortcuts and workarounds in software development usually happen for intentional reasons, such as for business deadlines and development complexity (Yli-Huumo et al., 2015a). Time-to-market and customer feedback are important factors for companies' success, and it is essential to deliver solutions on time (Lim et al., 2012). This is the reason why business stakeholders are often more focused on deadlines and customers than the actual quality of the software, which is more in the developers' interest area (Barney et al., 2008; Boehm, 2006). Therefore, strict deadlines may sometimes force the development team to create solutions with second-tier quality to meet the requirements within the deadlines set by the business stakeholders (Yli-Huumo et al., 2014). When TD starts to accumulate, it is often a safer and faster choice to take more TD with a quick and dirty solution, because there is a risk of breaking the product even more by modifying a complex part of the code base (Yli-Huumo et al., 2015a). Thus, code base complexity can force the company to take more TD intentionally, because the fixing of current TD would take too much time and money, while quick and dirty solutions are easier and faster to implement (Yli-Huumo et al., 2014).

TD can also occur unintentionally (McConnell, 2007). The reason for unintentional TD can be lack of competence, a need to upgrade existing technologies, or a customer or market -induced need for change. A coder may lack competence to develop an optimal solution. A development team may not be able to provide adequate instructions and coding standards for development, which reduces the quality of the solution. In legacy software, the old technology that is still in use can also be seen as unintentional TD. In these situations, a company sometimes has to start upgrading the technology to a newer version. It is also possible that changes coming from the market or a customer can turn the effort of the development team to a new direction. This means that previously developed parts need to be changed to make the product more suitable for the changing business needs. Fig. 1 shows a TD quadrant (Fowler, 2009) that identifies four categories of having TD for intentional and unintentional reasons (McConnell, 2007).

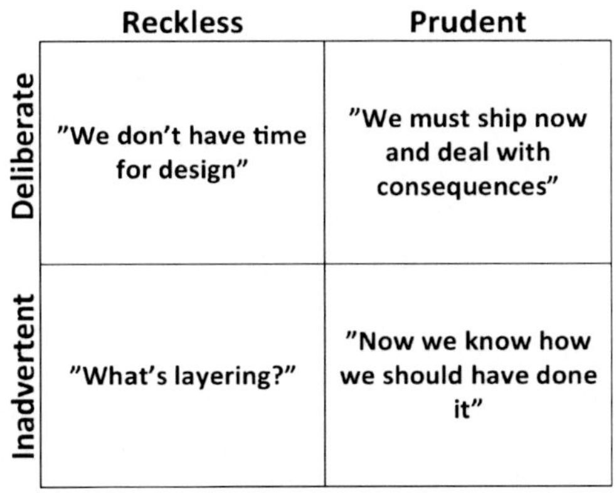

Figure 1: Technical debt quadrant (Fowler, 2009).

TD is often seen only as a negative concept in software development (Lim et al., 2012;Yli-Huumo et al., 2014). Software developers think that creating shortcuts and non-scalable solutions will increase the complexity within the code base (Yli-Huumo et al., 2014). When the code base starts to accumulate with too much TD that is not fixed afterwards, the development becomes more challenging, because the shortcuts are not designed to work well with other parts of the code base. Complexities in the code base start to reduce the overall quality and productivity goes down when new solutions and features must be implemented to the code base in debt (Yli-Huumo et al., 2015a).

Taking TD is never an optimal solution, and companies should avoid it when possible. However, actions that lead to TD can be beneficial to software companies, and in that sense TD can be seen only as a negative side effect. When taking TD, companies are able to speed up the release cycles to the customer, which can increase customer satisfaction and provide advantage in the market. Another benefit for companies is customer feedback (Yli-Huumo et al., 2015b). Companies are able to adjust the product and its business model based on faster customer feedback. This way the companies can identify and prevent both intentional and unintentional TD more efficiently, when customer feedback provides knowledge about the most important development needs in the software (ibid.). Therefore, while TD in the software is never a benefit, actions that incur TD into software can be beneficial to a software company in terms of acquiring business advantage and knowledge of customer and business needs.

Overall, the current conceptualizations of TD vary, and there is no clear, common definition. According to some scholars, TD should be associated only with intentional decisions happening in the code base, and messy code should not be counted as TD, while some think that old technologies in legacy software should also be counted as TD (Norton, 2009; Fowler, 2009). The addition of multiple terms related to shortcuts happening in other stages of the life cycle of software development also confuses the concept. In this study we focus on TD related to a badly structured architecture/code ("smelly code") and a code that violates coding guidelines. Even though concepts such as social debt (Tamburri et al., 2013) and people debt (Alves et al., 2014) describe similar phenomena of having shortcuts and non-optimal solutions in software development and organization, we believe that they should be categorized as *sources for TD* rather than as actual TD. Therefore, the definition of TD we use in this study is the following:

"A badly structured technical solution or architectural design in the system, incurred by either an intentional decision or an unintentional side effect, which causes omitted quality and productivity"

Our goal is to understand how software development teams manage intentional technical decisions when making compromises in software development. We also believe that unintentional TD is an essential part of software development. Our aim is also to identify how development teams try to prevent and reduce both intentional TD and unintentional TD.

Technical debt management

Technical debt management (TDM) is conducted to manage, prevent, measure and reduce technical debt (TD) during software development. TDM includes processes, techniques and tools that are used in software development. The current literature related to TDM has identified and developed some processes and tools (Li et al., 2015a). Managing technical debt (MTD) workshops have gathered multiple studies related to TD and TDM in the past years (Seaman et al., 2015). However, TDM is challenging to implement, and it is hard for managers and developers to estimate and identify what and how much TD the current system has, how it will change, and what effects it will have in the future (Li et al., 2015a). Power (2013) identifies seven main challenges surrounding

TDM: (1) agreeing what technical debt is; (2) quantifying technical debt; (3) visualizing technical debt; (4) tracking technical debt over time; (5) impact of neglecting technical debt over multiple releases; (6) identifying technical debt as a root cause of defects; and (7) understanding the cost of delay.

The reduction and repayment of TD are done by refactoring or rewriting the bad solutions (Codabux and Williams, 2013). Refactoring or rewriting can be seen as processes for "changing a software system in such a way that it does not alter the external behavior of the code yet improves its internal structure. It is a disciplined way to clean up code that minimizes the chances of introducing bugs. In essence when you refactor you are improving the design of the code after it has been written" (Fowler et al., 1999, p. 9). However, changing old solutions in the code is not easy, because improving the code base requires a significantly competent developer, and the company cannot just use all development time on refactoring or rewriting the solutions. Therefore, having some assisting approaches to know when and what refactoring is needed can be useful for development teams.

A portfolio approach for TDM has been suggested by Guo and Seaman (2011). The approach is widely used in the finance domain as a risk reduction strategy for investors, to determine the types and amounts of assets to be invested or divested. The core component of the proposed approach is a "technical debt list" (ibid.). The list contains TD "items", each of which represents an incomplete task that may cause problems in the future. Portfolio management could be adapted to manage TD, where the company would collect all the TD items to a list and use it to reduce TD and to conduct refactoring systematically. Li et al. (2015b) have also developed similar TD list management for architectural technical debt (ATD).

Unintentional TD caused by changes in the customer or market can be harder to manage and predict, because the development team cannot necessarily know these TDs in advance. However, the current literature has identified some practices to prevent unintentional TD. Implementing coding standards to the development process can prevent TD, when the developers have a cohesive way to produce a similar style code, which makes it readable and modifiable (Green and Ledgard, 2011). Code reviews can be used to check other developers' solutions before the release to catch possible TD issues in the design (Mantyla and Lassenius, 2009). Also simple practices in agile methodologies, such as the Definition of Done practice can reduce TD in the early stages of development (Davis, 2013).

An extensive mapping study of 49 primary studies has been recently conducted by Li et al. (2015a) to understand the current state of the art on TDM. The study identifies eight activities for TDM: (1) identification detects TD caused by intentional or unintentional technical decisions in a software system through specific techniques, such as static code analysis; (2) measurement quantifies the benefit and cost of known TD in a software system through estimation techniques, or estimates the level of the overall TD in a system; (3) prioritization ranks identify TD according to certain predefined rules to support deciding which TD items should be repaid first and which TD items can be tolerated until later releases; (4) prevention aims to prevent potential TD from being incurred; (5) monitoring watches the changes of the cost and benefit of unresolved TD over time; (6) repayment resolves or mitigates TD in a software system by techniques such as reengineering and refactoring; (7) representation/documentation provides a way to represent and codify TD in a uniform manner, addressing the concerns of particular stakeholders; and (8) communication makes identified TD visible to stakeholders so that it can be discussed and managed further.

Overall, the current understanding of TDM includes some ideas for processes, techniques and tools to manage TD. Even though the current literature has started to tackle and identify the concept and solutions of TDM, the problem is that there is a need for more empirical evidence from real-life software development (Li et al., 2015a).

Empirical studies on technical debt management in practice

There are few empirical studies on TDM. Guo et al. (2011) use a specific TDM framework to track down one delayed maintenance task in a real software project. Their TDM framework starts from the identification of a TD item, which then will be added to a TD list. After this, the TD item gets measured based on the principal and interest, which are based on estimates. Then, the TD item is ready for prioritization based on cost and benefit. With this framework, the authors have been able to track down and quantify TD items, and see the costs of delaying maintenance tasks. A similar approach has also been used by other researchers to identify and document TD issues in order to make TD easier to manage (Zazworka et al., 2013).

Klinger et al. (2011) interviewed four experienced software architects to understand how decision-making regarding TD was conducted in an enterprise environment. The results showed that the decisions related to TD issues were often informal and ad hoc, which led to a lack of tracking and quantifying the decisions and issues. The study also identified that there was a large communication gap between technical and business people as regards discussion about TD.

Different tools have been developed for TDM. The SQALE method (Letouzey, 2012; Letouzey and Ilkiewicz, 2012) has been developed for the purposes of identifying, estimating, analyzing, measuring, and monitoring TD in a software. DebtFlag (Holvitie and Leppänen, 2013) has been developed to capture, track and resolve TD in software projects. The SonarQube tool and its plugins have been applied in several studies to identify and measure TD from software (Al Mamun et al., 2014; Griffith et al., 2014). A set of other tools to support TD management were identified in the mapping study by Li et al. (2015a).

Most of the empirical studies of TDM take in consideration only few aspects of the eight TDM activities (Li et al., 2015a). A specific tool to identify and measure TD does not help in other activities, such as communication or prioritization. There is a clear need to know how TD should be managed from the organizational point of view. The mapping study by Li et al. (2015a) found a large number of different models, methods, practices, and tools in the literature for each separate TDM activity. However, there is no single solution that takes the whole problem of TDM into account. Therefore, a framework or model for TDM that combines all TDM activities is needed, both by researchers and practitioners, to understand all the aspects of TDM.

RESEARCH PROCESS

Research methodology

This study is qualitative, and it uses case study as the research methodology.

The definition by Yin describes a case study as *'an empirical inquiry that investigates a contemporary phenomenon within its real-life context, especially when the boundaries between phenomenon and context are not clearly evident'* (Yin, 2003, p. 13). As a research strategy, case study is used to contribute to our knowledge of individual, group, organizational, social, political, and related phenomena (ibid.). Therefore, case study has been a common research methodology especially in social sciences. However, the case study methodology has also been used in economics (ibid.), and it has become more popular in software engineering as well (Runeson and Höst, 2008).

Software development is carried out by individuals, groups and organizations, and therefore social and political questions are of importance for software development, which makes software engineering a multidisciplinary area where case study is a relevant approach (Runeson and Höst, 2008). There are multiple ways to study TD in software engineering. The research can investigate the written code itself, where the focus is on understanding how a badly structured code affects the other parts of the software. This type of research can be done with quantitative research methods, where the results show measurements on how for example performance or other quality attributes change depending on different structural possibilities in the code base. It is also possible to study TD from the organizational point of view. In an organizational TD study the research focuses on how various processes and practices are used in software-related TD. This type of a study can be performed with qualitative methods, which can, for example, produce results that show how people and processes affect the existence of TD in the software.

Since the aim of this study is to understand how software development teams conduct TDM to control and reduce TD, rather than what are the best possible code structures or architectural solutions, we believe that the case study methodology is an effective approach to understanding how people with different responsibilities working together in software development have organized TDM. The case study methodology makes it possible to examine the concept of TDM in real-life situations, to gather qualitative data, and to add to existing research related to TDM.

This study can be defined as an interpretive exploratory case study (Robson, 2002), as the goal of the study is to discover how software development teams have organized TDM, without a priori hypotheses. The purpose of an exploratory case study is to find out what is happening, seeking new insights and generating ideas and hypotheses for new research (ibid.). In addition, an interpretive case study aims at understanding phenomena through the participants' interpretation of their context (Runeson and Höst, 2008).

We decided to use the guidelines provided by Runeson and Höst (2008) to conduct the case study process. The case study process is divided into five main steps: (1) *case study design*: objectives are defined and the case study is planned; (2) *preparation for data collection:* procedures and protocols for data collection are defined; (3) *collecting evidence:* execution with data collection on the studied case; (4) *analysis of collected data*; and (5) *reporting* (ibid). Fig. 2 shows the research process used in the study, based on guidelines by Runeson and Höst (ibid.). The steps of the research process are discussed and explained in closer detail in the following subchapters.

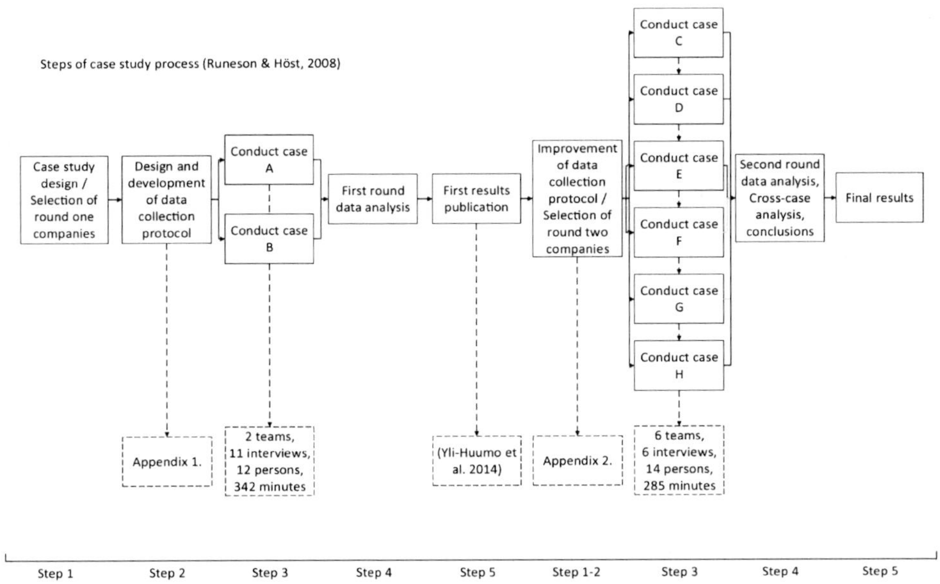

Figure 2: Case study research process used in this study.

Case study design and company selection

The design step of the case study process should contain the following elements: objective, the case, theory, research questions, methods, and selection strategy (Runeson and Höst, 2008; Robson, 2002). The design and development of the data collection protocol was started by examining the current literature related to TD and TDM. On the basis of the literature, we designed and developed a set of questions and topics for discussion to understand the reasons, effects and management related to TD. We decided to organize the interviews in two separate rounds. The reason for conducting the interviews in two rounds was that we wanted to understand the concept of TDM first through a smaller number of development teams to be able to adjust the interviews for the second round cases. This approach is similar to the theoretical sampling used in the grounded theory method (Strauss and Corbin, 1998), where the next data sample is chosen on the basis of an analysis of a previous sample, creating an iterative process for theory construction. In this case, we wanted to be able to modify our questions and topics based on the data received from the first round of interviews.

We decided to use *semi-structured interviews* (Charmaz, 2014) for data collection, which makes this research a flexible study (Runeson and Höst, 2008). Semi-structured interviews include a mixture of open-ended and specific questions, designed to elicit not only the information foreseen, but also unexpected types of information (Seaman, 1999). We thought that semi-structured interviews would provide us with good results, since the term 'technical debt' might be unfamiliar to the interviewees, and also taking in consideration the complexity in its definition, it was important to explain it carefully to create similar understanding between the interviewer and the interviewee. In addition, it was important to introduce all the aspects of TDM activities in the interviews, as it was highly possible that the interviewees would not have knowledge on their definition. Thus, the use of semi-structured interviews would make it possible for us to talk with the interviewees face to face, and in a case of misunderstanding, we would be able to explain the questions more precisely and ask more specific follow-up questions to identify answers to the research questions. A potential drawback of using semi-structured interviews can be the trustworthiness of the answers. However, we believe that there was no issue with the trustworthiness of the answers, since all the development teams in this study had expressed their interest in developing TDM in their organizations.

As our goal was to study TDM activities and their maturities, and we also assumed that the TDM activities would be used differently in teams within an organization, we decided to use the multiple-case study approach. Yin (2003) separates case studies to *holistic case studies* and *embedded case studies*. In holistic case studies the case is studied as a whole, while in embedded case studies multiple units of analysis are studied within a single case. This research fulfills the characteristics of a holistic case study, as our goal was to study each development team as a whole to understand the process related to TDM, instead of studying multiple units within one development team. The reason for studying multiple software development teams over one single team was gathering a broader amount of empirical data related to the research topic. We believed that studying several development teams would provide us with more information related to TDM, and comparing the results would help us understand what approaches were the most commonly used ones and why.

The selected case company is a large software supplier with around 5600 employees, currently operating in multiple countries in Europe. The company is a supplier of business software and business process solutions, outsourcing services, commerce solutions, and IT consultancy. It has currently about 340,000 customers. We studied eight software development teams in the organization. A summary of the software development teams and their roles in the organization is presented in Table 1. We selected the case company because its size, number of teams, and industry area, which made it very suitable for studying TD and its management. In addition, even though all the development teams were from the same organization, most of them were not working on the same product. Instead, most of the teams had their own product in development and a separate management, originating from the company's history of mergers and acquisitions. The company combines several product lines and includes teams coming from different backgrounds and cultures, but currently sharing the same organization. Therefore, we considered the company to be optimal for studying TDM activities in development teams.

Table 1: Summary of the case software development teams.

Team	Role of the team in the organization	Description
A	Development of a single product line.	The product provides a financial management solution as a cloud service. The current size of the team is 18. The whole development team is located in the same country. The development team uses a Scrum-like approach as the development methodology.
B	Development of a single product line.	The product is a SaaS-based project management solution for multi-organization projects. The current team size is 13, and the whole development team is located in the same country. The development team uses a Scrum-like approach as the development methodology.
C	Development of a platform infrastructure.	The platform infrastructure is used by the other development teams that develop the actual products for customers. The goal of the development team is to be a gateway between all the products and integrations within and outside the organization. The current team size is 12, and the development is distributed into several countries in Europe. As the development team works with other product lines in the organization, it uses Scrum and Kanban in parallel.
D	License integrations	The development team has two main responsibilities: (1) licensing and generating the invoices based on the usage, and granting the rights and accounting the usages for the services the customers have bought, and (2) integration of all the smaller systems that have been bought by the company. The current team size is 11, and the development is distributed into several countries in Europe. As the development team works with other product lines in the organization, it uses Scrum and Kanban in parallel. The development team also works closely with development team C.

E	Development of services and software for the organization.	The main responsibility of the development team is a platform from which all the services of macro segments are started and administrated. The current team size is 12. The development is also distributed into several countries in Europe. The development team uses a Scrum-like approach as the development methodology.
F	Development of a single product line in the organization.	The product is a web-based solution that allows approving invoices and expense claims online or mobile. The current team size is 8, and the whole team is located in the same country. The development team uses a Scrum-like approach as the development methodology.
G	Development of a single product line.	The product is a web-based solution for making budgeting and forecasting predictions for business monitoring and reporting. The software collects information on the company's financial records and other systems, as well as the form of real-time reports. The current team size is 8, and the development is distributed into several countries in Europe. The development team uses a Scrum-like approach as the development methodology.
H	Developing integration and security tasks for the organization.	The main goal of the development team is to handle integration and security tasks for the organization. The current team size is 17, and the development is distributed into several countries in Europe. The development team uses Scrum as the development methodology.

Data collection

The semi-structured interviews were conducted in two rounds between February 2014 and April 2015. The first round with two development teams located in Finland (Cases A and B) was started in February 2014, and it lasted until April 2014. We started the interviews by contacting the manager of the team. The manager of team A gave us also a referral to the manager in development team B. We conducted the first two interviews with the product line managers of teams A and B. After that we used the snowballing technique (Charmaz, 2014) to get referrals to other persons in the teams. As both development teams were located in Finland, we were able to travel physically to the offices and conduct all the interviews face to face. The total number of interviews was five in development team A and seven in development team B. In one of the interviews in team B (Interview ID B3), we interviewed two persons at the same time because of schedule constraints. The interview sheet for the first round interviews is enclosed in Appendix A.

The second round started in March 2015 and lasted until April 2015. Before starting the interviews, we decided to make changes to the interview structure for two reasons. The first reason was that the data gathered and analyzed in the previous round gave us new ideas for the interviews regarding TDM. In the first round interviews we identified a lack of TDM activities, which gave us an idea of focusing more on TDM. In the previous interviews, the focus was also on the effects and causes of TD. The analysis of causes and effects of TD is available as a separate publication by Yli-Huumo et al. (2014). The second reason for focusing more on TDM was the publication of the TDM mapping study by Li et al. (2015a). The mapping study identified eight TDM activities, which we considered as a good basic core for the inquiry on TDM. The results of the mapping study gave us new ideas for improving the interview structure more towards TDM activities. The updated structure for the second round interviews is shown in Appendix B.

The second round consisted of six software development teams located in various countries in Europe. The team manager of development team A gave us new referrals, which we used to contact the other six development teams. Because the teams were not located in Finland, we had to change the interview method from face-to-face interviews to online video calls. The interviews also changed from single person interviews to two-three person interviews. This was required because the time allowed for us was limited. The interviewees were usually one team manager and one software architect discussing the approaches to TDM. The risk of interviewing two or more people at the same time is that the interviewees would not necessarily be able to speak openly because of the presence of another interviewee. However, we noticed during the interviews that this was not the case, and all the six development teams were eager to talk about the problems with TD and TDM, and wanted to find possible solutions for improvements. In addition, we noticed that all the software architects had multiple years of experience with the software product, which was appreciated by the project managers involved. Therefore, we believe that the interviews were not disturbed by having multiple people present at the same time. Instead, we believe that the quality of the interviews was improved, since there was a common goal from both business and technical perspective to understand and improve TDM. The roles of the interviewees are shown in Table 2. When the interview engaged more than one person, this is referred to in the interview ID as E1a, E1b etc.

Table 2: Roles of the interviewees.

Interview ID.	Round	Team	Role(s)	Experience in the organization
A1	1	A	Software architect	6 years
A2	1	A	Software designer	1 year
A3	1	A	Project manager	4 years
A4	1	A	Software test engineer	1 year
A5	1	A	Product line manager	14 years
B1	1	B	Software architect	6 years
B2	1	B	Software developer	6 years
B3a	1	B	Product line manager	2 years
B3b	1	B	Software test engineer	4 years
B4	1	B	Software architect	5 years
B5	1	B	Software developer	1 year
B6	1	B	User interface designer	1 year
C1a	2	C	Team manager	5 years
C1b	2	C	Software architect	17 years
C1c	2	C	Software architect	3 years
D1	2	D	Software architect	7 years
E1a	2	E	Team manager	15 years
E1b	2	E	Software architect	8 years
E1c	2	E	Software architect	5 years
F1a	2	F	Team manager	4 years

F1b	2	F	Software architect	9 years
G1a	2	G	Team manager	1 year
G1b	2	G	Software architect	4 years
H1a	2	H	Team manager	3 years
H2b	2	H	Software architect	3 years

Data coding and analysis

In exploratory case studies, the technique for the analysis of qualitative data is hypothesis generation (Seaman, 1999). As we did not have any priori hypotheses for this study, our goal was to use the techniques for data coding and analysis of qualitative data to find hypotheses from the collected data and interviews. The techniques for data analysis used in exploratory case studies are *constant comparisons* and *cross-case analysis* (Seaman, 1999).

Fig. 3 gives an overview of the data coding and analysis processes conducted in this study. The data coding and analysis were completed in various steps, guided by the work of Robson (2002). Overall, we conducted a total of 17 interviews with 25 persons related to eight studied cases, and had 627 minutes of audio-recorded data. When all the interviews were conducted, we began the data transcription phase. The first round interviews were transcribed by the authors, and the second round interviews by a hired person with English language proficiency. The reason for the authors to transcribe the first round interviews was that the interviews were conducted in the Finnish language. The authors transcribed and translated the first round interviews to the English language to make the coding and analysis stage easier, because there would be only one main language in use in the study. All the second round interviews were conducted in English. During the interviews we were also able to gather some additional documentation data. In one of the interviews we received a PowerPoint presentation related to the TDM activity the team was currently conducting.

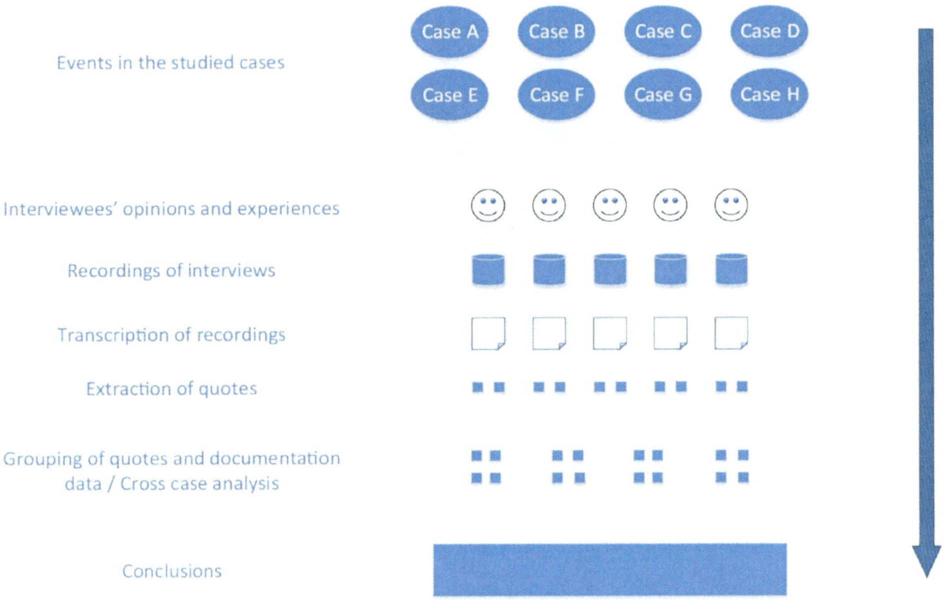

Figure 3: The coding and analysis process.

After all the data was transcribed, we started the data coding and analysis stage. The total word count of transcriptions in Word was 73 955. We used a tool specialized for qualitative data coding and analysis, Atlas.ti. In data coding, one code is usually assigned to many pieces of text, and one piece of text can be assigned more than one code. The codes can form a hierarchy of codes and sub-codes (Robson, 2002). Our data coding stage followed the top-down approach, because the categories were derived from the mapping study by Li et al. (2015a), which identifies eight activities for TDM. The categories used in the data coding were *TD repayment, TD representation/documentation, TD identification, TD prioritization, TD measurement, TD monitoring, TD communication, and TD prevention.* Table 3 shows an example of the data coding process with Atlas.ti, where the interviews are used to extract quotations to the identified categories. We believe that using the top-down approach in the data coding was an effective way to understand how every TD activity was approached in every development team, which helped us to draw conclusions and understand the TDM process.

Table 3: Example of the data coding process.

Interview transcripts	Categories
"We have Epics, and we have some kind of goal, so 20% of developer time to be in internal quality."	TD repayment
"We have quite often security reviews, and maybe some technical debt can come from security"	TD prevention
"We can measure also how much time we spend on this slice of the backlog. For example, on internal quality as the whole in the team, think, spend 218 h, and we see people, some our system architects, some our developers, some our QA testers and so on. So, we have this console and having this an objective 20% of developer time, we can check if we spend that time, that budget on not."	TD monitoring, TD measurement, TD repayment
"Also as a team we have some KPIs. In the team, we are part of product unit, R&D department for ERPs and other product-related. And then we as a team define KPIs to measure, and for example, we have one key control on technical debt."	TD communication, TD measurement, TD monitoring
"And as a team, we have kind of demo, some kind of retrospective on monthly basis, then we do all the numbers and then discuss it: this strength is good, this number is too low, what to do, and then we put on backlog actions."	TD communication, TD monitoring, TD documentation

When all the quotations were extracted and identified to the specific categories, we analyzed every case independently and drew a conclusion on the process used for TDM in each case. When we had a complete view on every case, we started a cross-case analysis to find out the similarities and differences between the cases.

RESULTS

Case A

TD repayment with refactoring and rewriting was based on the general development backlog, where some of the code base improvement issues could be found. However, we were not able to find any repayment strategy for the TD that was incurred during the development. The developers in the team mentioned that it is sometimes impossible to get time to refactor the solutions that were developed previously with shortcuts. The reason was that new features were already waiting in the next sprint's development backlog that were prioritized higher than technical improvements in the code base. Therefore, TD repayment with refactoring or rewriting was mostly done unofficially during the actual development time that was reserved for new features. Sometimes this refactoring was not even mentioned to the management. The team management had adopted a practice where every Friday was dedicated to bug fixing. However, the developers felt that it was mostly dedicated to fixing only bugs, instead of conducting architectural-level refactoring or rewriting.

TD representation/documentation was not systematically conducted by the development team. The development team did not have a separate TD backlog to document TD items either. When a developer identified a possible TD in the code base, there was no clear process or guideline on how to document it to the management system. One of the developers mentioned that the team used the JIRA management system, where it is possible to create tickets for issues found during the development. However, this was not always done by the developers, which resulted in situations where some TD remained undocumented and was kept in the notes of the developers, and even sometimes forgotten.

TD identification was mainly conducted during the development, when a developer noticed a problematic area in the code base, which then sometimes resulted in fixing the case in the management system. TD identification was also conducted by the system architects and team managers, who sometimes analyzed the code base to find what should be changed to improve the quality and maintainability, and added some refactoring tasks to the development backlog.

TD prioritization was mostly done on a hunch. When a TD issue was raised in the management system, the development team would discuss the importance of that specific case and give it prioritization. The most important factors taken in consideration when deciding issues to be refactored were the scalability and business value of that specific feature.

TD measurement and *TD monitoring* were not conducted by the team manager or the software architects. The reason for not measuring or monitoring TD was the fact that the development team did not have any clear process for documenting TD items, which meant that the team management did not have the possibility to gather or analyze any clear data. Only estimations of TD were based on current knowledge about the code base and issues in it.

TD communication was structured well and the development team members understood the concept of TD. The team manager had a good technical knowledge background, which helped the developers and software architects to communicate and discuss about the possible TD issues that had occurred during the development. This was the reason why the development team would sometimes get more time to fix and repay TD issues that had been bothering them in the actual development. The team manager would also act as a filter between the business team and the development team. When the business stakeholders gave tasks that were impossible to develop within the given deadlines, the team manager would explain the situation to the business managers, which sometimes gave more space to the developers.

TD prevention was done with coding standards and code review practices. The development team had taken in use some level of coding standards with coding books and instruction videos to show the developers what kind of coding was expected to be developed. Code reviews were sometimes conducted by two software architects, but it was not mandatory to check every newly developed code. As supervision of TD prevention was not always conducted, one of the developers mentioned that sometimes the developers would just use the old code and copy it to the other parts of the code base, which could be risky.

Overall, the TDM strategy in development team A was not organized as a systematic process. The development team did not have any clear TD documentation or TD repayment process to gather TD issues, and it was often organized unofficially. This was the reason why it was also impossible for the team management to monitor or measure TD. However, the development team had a good basis for starting TDM, because the communication was active regarding TD, and the team manager was considered to have good technical knowledge, which helped the development team to deal with TD. The development team also had good TD prevention practices in use, even though their actual use was not confirmed.

Case B

TD repayment in development team B was mostly considered as part of the normal work during the development. In a situation where a developer identified a small refactoring case, there was no need to create a separate issue out of it. In a situation where the refactoring case was bigger, the development team would organize a discussion with the team manager and software architect to discuss the next steps and whether there was a need to conduct refactoring or rewriting of that specific solution. The team management had also organized a practice where one day of the week was dedicated to fixing bugs and making small refactoring.

TD representation/documentation was not currently a systematic practice within the development team. When a developer decided to take a shortcut during the development, there was no mandatory process defined on how it would get stored and documented in the JIRA project management system that was used. We identified situations where the developers might have created JIRA tickets to the management system, but also situations where they were just left in the coder's own notes. The developers also did not always inform the management about what shortcuts were made. The team manager and the software architect would sometimes add some TD issues to the development backlog, when the issues were raised during the development.

TD identification was mainly done during the development by the developers. When the code base was developed, the developers would identify the refactoring needs, when the currently developed part was extremely complex and hard to develop. Sometimes also the software architects would go through the code and try to identify possible places, especially in the architecture, where refactoring was needed.

TD prioritization was often based on a hunch and previous experience with the code base. However, prioritization would get done according to the location of the issue. If the issue was in the core of the code base, depending on several other places, it would get prioritized as highest. After this, issues in the business logic and user interface were prioritized under it.

TD measurement and TD monitoring were not currently done by the team management. The reason was that it was at the moment impossible for the team manager and software architects to know what TD the software currently had, because it was not properly documented anywhere. The team management did not have any specific tools in use to measure TD, either. This was the reason why there were no accurate measurements or monitoring to see the current status of TD.

TD communication was structured well in the development team. The team manager had wide technical knowledge, which helped TD communication between the management and the developers. This also helped the development team in situations where the business stakeholders gave impossible deadlines to work with, because the team manager would explain the issues of possible TD to the business management. However, the development team expressed a problem in communication, as the development of new features was always prioritized the highest, and the code base improvements were not done before them.

TD prevention was done with coding standards and code reviews. However, the team manager mentioned that they were not on a good level at the moment, and there was a need for improvement. The current coding standards did not fulfill the needed requirements, and the development team did not always follow them. Also, the code reviews were conducted by the software architects, but it was not always possible to go through all the developed code, as it was not prioritized enough.

Overall, the TDM strategy in development team B was similar to Case A. There was currently no mandatory process used to document TD issues in the JIRA system, and the development team did not have a special TD backlog in use. Refactoring was conducted mostly unofficially during the development, and there was no systematic process to repay TD in certain periods. Similar to Case A, this was the reason why it was extremely difficult for the team management to measure and monitor TD. However, the idea of TDM was understood by the management, and they were eager to find improvements. This is why TD communication was active within the development team, which gave more space for them to work on some TD issues.

Case C

TD repayment was identified as an essential part of software development by the software architect, and the management of the team had realized that it should be a part of the development process. Development team C used the Kanban methodology and JIRA tool to manage the software project. In the Kanban table, the team management had assigned 20% of the development time specifically to improving internal quality. Internal quality was divided into five main parts: refactoring, test automation, DevOps, platform security, and performance. The development team used these five internal quality factors to assign issues to if something needed to be refactored, rewritten or redesigned. The development team identified TD as an important key performance indicator within internal quality. If a person in the development team saw a bigger need for refactoring, he/she created an issue in JIRA under internal quality, which then was included in the actual development backlog after a discussion with the team management and software architects. Smaller refactoring cases where just done during the development without mentioning them to the management.

TD representation/documentation was not always done systematically by the developers. The development team did not have any mandatory guidelines for the developers for representation and documentation of TD. In a case where a developer created or founded a TD issue, there was no clear process of how to document it systematically afterwards. Instead, the developer may have sometimes created a JIRA issue ticket for refactoring, and it could be found in the internal quality section, or in some cases it would not get documented. The internal quality section in this case was used as a TD backlog. The management felt that this should be improved a lot in the future and there should be clearer guidelines for systematical documentation of TD.

TD identification was conducted mainly by the software architects. The two software architects were given responsibility by the management to identify TD in the code base. Therefore, the software architects usually went through the code base to understand and identify possible items to refactor and improve, which were then added to the internal quality issues. The identification was often done just by going through the code base manually, and trying to understand what parts of the code base were the most complex ones. Part of the identification was conducted with the SonarQube tool, but the architects mentioned that it was not necessarily the best way to identify all TD, because it does not take deep and complex architectural issues into consideration. For example, with SonarQube the software architects were able to find issues related to single line problems or code violations, but it could not detect some complicated business logic issues, which was considered as a real technical problem. Therefore, the identification was seen more as the responsibility of the people and processes. The team manager and software architects also mentioned that the developers were not currently involved heavily in the TD identification process, and hoped that they would start to identify more TD issues in the future.

TD prioritization responsibility was given to the software architects. The prioritization of TD issues was usually done on a hunch and previous experience of the code base. The development team did not have any systematic way to give estimations or numbers to prioritize TD issues. One of the software architects mentioned that sometimes they would take into consideration issues like how heavily the feature was currently used or whether a lot of new features were expected to come to that area in the future. The team manager mentioned that he trusted people's opinions more than numbers when making decisions about refactoring.

TD measurement was done by one of the software architects, who used SonarQube to measure TD. The SonarQube tool gave values of TD as automated test coverage and violations in the code. The software architect used these two measurements to estimate the current TD monthly. However, the software architect responsible for the measurement thought that using only SonarQube to have measures of automated test coverage and violations in the code base cannot be the only good way to measure the actual TD. The problem was that SonarQube only identifies minor TD issues, such as issues in the code, but not real problems in the architecture. This was the reason why it was hard to generate refactoring issues from the SonarQube tool to the internal quality backlog.

TD monitoring was done by using data gathered from the JIRA tool, which gave the management the possibility to estimate and follow how much time had been spent on internal quality compared to the overall development time in a certain period, and whether it was aligned with the agreed 20% rule. The software architect also used data from SonarQube to monitor the current status of TD monthly, and it was analyzed and reported to the management, to show whether TD was increased or decreased. The combined data from JIRA and SonarQube was used to monitor how TD was chancing.

TD communication was an important area of discussions between the team management, software architects and developers. The team manager worked closely with the software architects, which helped in communicating about issues related to internal quality and TD. This way the development team was able to reduce issues related to internal quality and TD, instead of using the development time to create only new features with business value. The software architect also often discussed with the developers about issues related to TD.

TD prevention was conducted sometimes with coding standards and code reviews. The team used Java coding standards as a recommendation to developers to produce similar code. Both software architects also sometimes reviewed the code to catch bad designs. However, these were only used as recommendations, and it was revealed that in reality the coding standards were not always followed or code reviews conducted.

Overall, the strategy for conducting TDM was structured well in development team C. The idea to use 20% of the development time to improve the code base and refactor architectural issues was a good strategy to reduce TD systematically in the software. Also, the measurement and monitoring with the JIRA and SonarQube tools gave the management some level of estimations about the current status of TD in the software. The issues with TDM in development team C were TD documentation and TD prevention. Even though the TDM structure was well-designed to repay TD systematically, the development team did not have a proper documentation practice in use. When the developers took or found TD issues, they were not always reported or documented, which made it hard for the software architects to understand the status of the current TD issues. TD prevention with coding standards and code reviews was also lacking and considered a big problem by the management.

Case D

TD repayment was conducted, similarly to Case C, by assigning 20% of the total development time to reduce TD issues in the software. The time for improvements was mostly used for additions of automated tests and unit tests. The software architect of the team felt that they could reduce TD the most, because it prevents TD from occurring in the software. If a need for refactoring was found during the development, it was assigned to the 20% internal quality section in the JIRA management system that was used in the team. The 20% rule was also used for bigger refactoring and rewriting issues to remove bad designs from the code base. The development team had a two-month release cycle, where the last two weeks were dedicated to the stabilization of the code base. During the two weeks, the development team would discuss current TD issues and what should be refactored in the next two months' iteration. The software architect also had the authority to use the JIRA system to see internal quality issues, and make decisions on what should be refactored in the next iteration. The goal was to fulfill the 20% rule in every iteration. Sometimes only for example 10% was required to be used on internal quality, because there may have been a need for new features with important business value. However, the team manager may have added 25% to the next iteration after that, to keep the average on the agreed 20%. The refactoring or rewriting of small TD issues was conducted during the development, and it was not necessary to mention them to the management or report to the system.

TD representation/documentation was done to the JIRA management system. When a member of the development team saw a possibility to have a refactoring case, the instruction was to create an issue to the system about it. In addition, if a developer needed to take an intentional shortcut during the development, it was also instructed to be reported in the system. In this case, the internal quality in the JIRA system worked as a TD backlog.

TD identification was done mostly by the software architect reviewing the code base manually or with a tool. The tool used for identification was SonarQube. The software architect ran the SonarQube tool every night when the new version of the software was out and used it to gather statistics about TD. If the tool reported any major issues, it was the responsibility of the software architect to report and go through every critical issue and try to fix them before the end of the iteration.

TD prioritization was done by simple low, medium, high, and blocker scales. High and blocker TD issues were repaid immediately or in the next iteration. Medium TD issues were also repaid in the next iteration or the one after that. Low TD issues did not usually get repaid ever, because the backlog was usually flooded with them. The software architect felt that fixing low priority issues would not bring any value to the software. The management and software architect also assigned story points to TD issues, based on the Fibonacci scale. For example a medium case was usually assigned 5 or 8 points, while high or severe cases got 13 or 21 points. The management and software architect responsible for prioritization did not use any specific calculations to create these prioritizations. The decision about a single prioritization was mostly based on a hunch and the experience of the software architect with the code base.

TD measurement was conducted with the SonarQube tool by the software architect. The results of the SonarQube were used to have a measurement of the current TD. The team manager also used Fibonacci scale prioritization to measure the velocity in the development in order to understand how much the development team could repay TD in the next iteration. The management and software architect felt that these two measurements for TD could be used to have good TD estimation.

TD monitoring was conducted with the JIRA and SonarQube tools. The management was able to use JIRA as a tool to monitor the development time for various tasks on either new features or TD reduction. The management used this information to generate reports at the end of each iteration. The software architect also tried to use the data received from the SonarQube tool constantly as a way of monitoring TD.

TD communication was performed between the team manager and software architect, who discussed the importance of TD repayment. The team manager initiated the discussion at the beginning of each iteration to discuss and list things that would need to be done in the next iteration. The software architect mentioned that the development team was currently in a lucky situation, because the team manager understood the concept of TD and was eager to help the development team in dealing with it. Even though the current team manager was not described as someone who took part actively in technical decisions, she still understood the importance of TD and the fact that software quality would be an important factor in the long term, which gave increased visibility to the effort of reducing TD.

TD prevention was conducted with coding standards and code reviews. The development team had created a rule that nobody could not commit anything to the code base before another developer had reviewed it and it fulfilled the standards of the Definition of Done. Of course in reality this meant that if a developer changed a minimum amount of code, it would not be necessary to be reviewed, but in a case where there was a risk of breaking the software, a review was mandatory. In the most challenging cases, more than one review was needed. Also a discussion with the whole team was organized to understand, learn and find the best solution. The software architect mentioned that even though this rule was good to have, it was not always followed very strictly.

The overall TDM strategy in software development team D was constructed well. The management had a clear vision and understanding of the fact that 20% of the development would be used for TD repayment. This was compounded with the JIRA and SonarQube tools that were used to document, measure, monitor, and identify TD. The development team had also well-conducted rules in code reviews and standards to prevent TD.

Case E

TD repayment and improvement decisions of the code base were created on the basis of a stakeholders' meeting once a month. The manager of the development team would make a suggestion in the stakeholders' meeting on how much time would be needed to repay TD and improve the code base in the next month. The manager of the team mentioned that for example in the three previous months, the development team had made an agreement with the stakeholders that one third of development was assigned to repaying TD. However, the problem with the repayment of TD was that even if the development team got the time agreed to refactor or rewrite issues, in reality it was not possible to do so, because the new features would always take more time to complete than estimated, which took away time that was reserved for TD repayment.

TD representation/documentation was done by creating a backlog approach for TD issues. When a development team member made a decision to create a shortcut to some solution, or identified a need to refactor old technology or bad design, it was documented in a separate 'technical debt backlog' in the JIRA management system. This process was used by the development team to make TD more visible in the development process. As the nature of the development team was to act as a platform for other product lines in the organization, the backlog was also used to communicate about TD issues in the platform with other development teams. The development team was able to present the backlog to the other development teams with information about possible issues in the future where TD would be most disturbing.

TD identification was mostly done by the software architects, who spent a lot of time with the code base, trying to identify possible improvements regarding TD. The development team did not have any special tool to identify TD in the code base, and it was mostly done by just "smelling the code".

TD prioritization was done by the team manager and software architects. However, the prioritization process was described as not well constructed. The team manager and software architects mentioned that they would categorize the first TD according to its type. Issue types were usually related to refactoring, security and performance. After this, the actual prioritization was mostly done at a hunch, based on opinions and experience with the issue. The most important factors taken in consideration when making a prioritization were often related to how TD would affect the customer and future projects. Also security and performance were mentioned to be important when deciding on the most important refactoring cases.

TD measurement and *TD monitoring* were mainly done with information that was available in the JIRA issues. The TD backlog in JIRA was used to gather some statistics and to estimate the current status regarding TD in the platform. The manager and software architects used some basic information in the backlog to monitor and measure how much TD the platform had by calculating the number of issues and prioritizing them according to their importance, to make refactoring.

TD communication was done mainly in the stakeholders' meeting. The management and software architects gave suggestions to the stakeholders in the meeting about the currently highest prioritized TD and why it should be important to repay and refactor as soon as possible to prevent future issues with it.

TD prevention was done with code guidelines and code reviews. The developed code was always checked with a tool called StyleCop to ensure that it was written according to the guidelines. It was also mentioned that the developers did sometimes do code reviews, but this was not described as mandatory. If a developer wanted someone to check the code, he/she could ask someone to do it. However, code reviews were instructed to be conducted if the developed code was done in a section of the code base that was known to be extremely complex. The management mentioned that the reason for not conducting code reviews in all developed codes was that most of the developers' time was assigned to the development of new features, and there was no time to have all codes reviewed. Overall, the TDM strategy in development team E was mainly focused on the documentation of current TD issues, and trying to find time to refactor on the basis of stakeholders' meeting once a month. The management did not currently have any systematic way to identify, measure or monitor TD. The management also mentioned that TD prevention was currently not the most effective, and more time should be reserved to it.

Case F

TD repayment was not done in any organized process. One of the software architects mentioned that in the current process, the repayment of a TD issue was usually started only when the issue started to be highly problematic for the development team to handle and there was no other way than just to refactor, rewrite or redesign it. If a developer noticed a need for refactoring, it was often taken care of by the developer, without any actual systematic repayment process.

TD representation/documentation was not a part of the development process. Sometimes the software architects of the team would add some major issues identified in the code analysis tool which would need to be addressed, to the main backlog. In a case where a developer took a shortcut or noticed a need for refactoring, it was often just left in that developer's memory, and may be documented somewhere.

TD identification was conducted by using SonarQube, CheckStyle and FindBugs as tools to analyze the code base to find possible TD. The software architect used the data gathered with the tools to understand the current status of TD in the software. However, similarly to Case C, opinions about the actual data acquired from the tool varied. The issue was that the tools did not necessarily give the needed information about TD in deep architectural structures of the code base. However, the software architects took the most critical issues identified by SonarQube and tried always to fix them. The common opinion was that the actual identification was done during the actual development, and the development team had some self-assessment cases to identify TD issues. SonarQube was not advised to be used by the developers, so identification with a tool was done mostly by the software architects.

TD prioritization was mostly done at a hunch and the software architects used their previous experience with the code base as the starting point when prioritizing TD issues.

TD measurement and *TD monitoring* were not conducted by the team manager and software architects, even though they used the SonarQube tool actively to identify TD. The reason was that SonarQube did not give valuable numbers for actual measuring of the real TD. The real metrics used were the actual global number of TD issues in JIRA, which was used to measure the current TD.

TD communication was seen as a problem during development. The development team felt that communication about TD to the business people in the organization was difficult. The software architects also felt that the development team did not currently discuss issues related to TD with the team management or software architects.

TD prevention was not conducted at a good level within the development team. The development team had set up some standards with the SonarQube, CheckStyle and FindBugs tools. The development team mentioned that they did not currently have much stuff related to coding standards or code reviews. However, the team manager mentioned that they were currently developing a definition of the done standard to improve TD prevention in the future. The future Definition of Done should comprise at least code reviews, unit tests, and errors found by SonarQube.

Overall, the TDM strategy in development team F was not an important issue within the development process. The development team did not currently have any systematic way to document, monitor or measure TD items. Repayment was often based on a hunch and was conducted when some TD issues started to grow too large, and the only way was to refactor or rewrite the solution. The reason why the management thought that implementing TD processes to development would be challenging was that if the team conducted constant identification and repayment of TD, it would not be cost-beneficial to the organization.

Case G

The *TD repayment* process was often started on the basis of a feeling that something should be improved. When the software architect or a developer noticed that there was a need for bigger refactoring or rewriting in the code base, it was mentioned to the management. After this the management would organize a discussion about the issue, where the development team would estimate the effort to fix the issue. The team manager then used these estimations to insert TD issues into the development backlog in future sprints. However, sometimes these issues were forgotten in the JIRA system, and they were never repaid. The team manager mentioned that currently the development team did not spend very much time on TD repayment. Smaller TD issues were just fixed during the actual coding.

TD representation/documentation was not done in any separate backlog. When a developer took a shortcut during the development, he/she would sometimes create a ticket to the JIRA management system, where information about quick solutions could be found. However, the software architect mentioned that the development team often took shortcuts that were not mentioned to the management, and this information was only stored in that developer's own notes.

TD identification was not conducted by the development team. The manager mentioned that there was currently no systematic way to review the code and identify possible TD issues. Identification would only start when there was a clear issue and an urgent need for a fix. The software architect mentioned that he had created a memo in the development team's WIKI page about TD issues he had identified and thought should be fixed.

TD prioritization was done by the manager and software architect. The prioritization process did not have any specific calculation to rate TD items, it was mainly done at a hunch. The team manager mentioned that when making the decision on what to repay next, factors like time, functionality, further maintenance, scalability, business value, and future plans with that feature were taken into consideration and used to give priority to various issues.

TD measurement and *TD monitoring* were not currently done by the manager or the software architect of the team.

TD communication was not described as active within the development team. The software architect mentioned that the development team was not currently talking about these kinds of issues during sprints. Also communication with the business owners was described as challenging, and usually the priorities they gave consisted only of development of new features, and not improvement of the code base.

TD prevention was not done by the development team. The software architect mentioned that the development process did not currently contain any coding standards or code reviews, and this was a huge problem. Everyone just used their own style of developing, and there was no consistency. The team had tried to use coding standards and code reviews before, but the usage was stopped because it was seen as time consuming. Another big problem mentioned by the software architect was that the development team did not have any proper Definition of Done to the development. The only Definition of Done was that when the solution was in production, it was considered to be ready. This was why a lot of bad solutions were created in the code base.

Overall, the TDM strategy in development team G was not organized systematically. It seemed that the management and developers did not have an explicit process of how to repay TD on a clear basis, and the development time was always put towards new feature development. The management did not have any way to measure or monitor TD, because the development team did not have any definite process to identify and document it. The development team was also lacking in the prevention of TD, by not having any coding standards or reviewing of the developed code.

Case H

TD repayment was organized systematically to conduct refactoring during the software project. The management had decided to use a certain number of days each month for the improvement of code quality. In every month, two days were assigned for unit testing, where the developer unit tested every code of their own that had been created in the last month. Also, one more extra day of the month was dedicated to 'your review day', where every developer's code was reviewed by another developer. Also four to five days a month were dedicated for ′your development day', where the goal was to improve the quality of the code base. In case a developer needed to take an intentional shortcut during the development, he/she was guided to create a JIRA issue, which would be fixed in the next sprint.

TD representation/documentation was done by using a backlog approach to document all possible TD items happening during the development. When a developer took a shortcut during the development, it was issued as a JIRA ticket to the system, where the manager and the team could follow the possible improvement needs. *TD identification* was conducted during the continuous integration process. If the development team noticed that some part of the software needed technical improvement, it was something that should be focused on.

TD prioritization was based on story points that were assigned by the team management and software architect. If there was need for a big change in the code base, it would be prioritized higher. The prioritization was mostly based on a hunch and previous knowledge of the issue and that certain area of the code base. The team used figures from SonarQube and the opinions of project stakeholders to make the decision on what TD would be prioritized as the most important to repay.

TD measurement was done with the SonarQube tool. The team manager and software architect ran the code base with SonarQube and were able to measure the amount of TD every month. The results from the tool were compared to the standards and performance that the management had set up. This way the team manager was able to measure whether TD had increased or decreased during the previous month.

TD monitoring was based on JIRA and SonarQube information. The team manager felt that information from SonarQube was important information to monitor, to know how healthy the software was at any point of time. He also admitted that SonarQube did not necessarily offer information about big architectural issues, but still thought that it was valuable information to have.

TD communication was active between the management and the development team. If a developer identified a TD issue, it was discussed with the software architect, and the decision to allocate time for it was often granted.

TD prevention was conducted by creating a strict definition of the done process to every code that was developed. When a developer created or changed something in the code base, it was guided to be tested first locally in the developer's own machine. After this, the code went to the acceptance environment, where all the other components were connected to the system. When the code was tested and verified in the acceptance environment, it would go to the staging environment, where it would go through automated test cases. Finally, if no bugs or issues were found, the code would go to the production environment. In case an issue was found, the code would go back to the developer, who had to refactor or rewrite it.

Overall, the TDM strategy in development team H was organized systematically. The team manager described the process as continuous refactoring, where the goal was to keep the overall quality of the code base always on an acceptable level. This was conducted by having a continuous TD repayment and TD prevention strategy that was compounded with TD monitoring and TD measurement by the team management.

SUMMARY OF THE CASES

A summary of the cases is shown in Table 4. In TD repayment, all the development teams used either refactoring, rewriting or redesigning as the main process to repay TD issues. TD repayment was done during normal development and consisted of only small repayment cases or TD repayment that was done from issues assigned to the actual development backlog. Some of the development teams (Cases C, D, E, and H) had a systematic strategy to TD repayment, by assigning a certain amount of development time every month to improving code quality by refactoring or rewriting the solutions. We also identified teams (A, B, F, and G) that often started TD repayment when the TD issue started to become a problem and there was free time allocated to it.

Table 4: Summary of the cases.

Case/ TDM	TD repayment	TD identification	TD measurement	TD monitoring	TD prioritization	TD	TD prevention	TD
activity						**representation/**		**representation/**
						communication		**documentation**
Case A	Issues from the general development backlog. Refactoring during normal development.	Team manager and software architects identifying TD manually. Mostly during normal development.	No measurement	No monitoring. Sometimes from JIRA issues.	Mostly based on a hunch. Business value and scalability taken into consideration.	Communication structure good with highly technical team manager.	Some coding standards and code reviews.	Some issues to JIRA. No separate backlog.
Case B	Issues from the general development backlog. Refactoring during normal development.	Team manager and software architects identifying TD manually. Mostly during normal development.	No measurement	No monitoring. Sometimes from JIRA issues.	Mostly based on a hunch. The number of places affected by the change taken into consideration	Communication structure good with highly technical team manager.	Some coding standards and code reviews.	Some issues to JIRA. No separate backlog.

Case C	20% of development assigned to improving the code base. Refactoring during normal development.	Software architects identifying TD manually. Identification with SonarQube tool.	Team manager measuring from JIRA, software architects from SonarQube statistics.	Monitoring with JIRA and SonarQube.	Mostly based on a hunch. Sometimes taken into consideration how much the feature was used now and would be used in the future.	Communication about TD active within the whole development team.	Some coding standards and code reviews. Not always conducted.	Internal quality backlog on JIRA. Not used systematically by the developers.
Case D	20% of development assigned to improving the code base. Refactoring during normal development.	Software architects identifying TD manually. Identification with the SonarQube tool.	Team manager measuring from JIRA, software architects from SonarQube statistics.	Monitoring with JIRA and SonarQube.	Mostly based on a hunch. The management using low/medium/ high/ blocker - story points -Fibonacci scale	Communication about TD active within the whole development team.	Code reviews and coding standards. Definition of the done standard.	Internal quality backlog. Developers using systematically.

Case E	Decided once a month in a meeting. Refactoring during normal development.	Software architects identifying TD manually.	Measurement from the TD backlog.	Monitoring TD backlog.	Mostly based on a hunch. Taken in consideration how much it would affect the customer and future projects.	Communication in stakeholders' meeting monthly.	Some coding standards and code reviews. Not always conducted.	Separate backlog for TD items.
Case F	Refactoring only when TD became a huge problem.	Identification with tools (SonarQube, CheckStyle, FindBugs). Mostly during normal development.	Measurement from JIRA issues.	Monitoring with JIRA.	Mostly based on a hunch.	Communication currently challenging with business people.	Some minor coding standards and reviews of the used tools.	Some issues in JIRA. No separate backlog.

Case G	Refactoring only when TD became a huge problem.	Identification rarely done.	No measurement.	No monitoring.	Mostly based on a hunch. Time, functionality, further maintenance, scalability, business value and future plans taken into consideration.	Current communication of TD lacking.	No coding standards or code reviews.	Some issues in JIRA. No separate backlog.
Case H	Number of days in a month assigned for improvement.	Mostly during normal development. Identification with the SonarQube tool.	Team manager measuring from JIRA and SonarQube statistics.	Monitoring with JIRA and SonarQube.	Mostly based on a hunch. Story points based on how important the issue was.	Communication active within the development team and stakeholders.	Definition of Done to ensure code quality.	Issues reported to JIRA.

TD identification was done during the development or with tools. In many cases, TD identification was done during the development, when a developer or software architect noticed a problem with a solution during normal development or analysis of the code base. Sometimes these identifications would happen accidentally during the development, or a software architect would spend some time with the code base to identify if there was anything important to refactor. In some cases (C, D, H) the SonarQube tool was used for TD identification.

Most of the development teams (A, B, E, F, and G) did not have or did not know a good way for TD measurement. Some of these teams (E, F) mentioned that the only TD measurement information they had was the JIRA management tool, where there was a possibility to measure and calculate how many TD issues had been assigned to the system. Some of the development teams (C, D, and G) used the SonarQube tool to measure TD in the software.

Some teams (A, B, E, F, and G) did not have systematic TD monitoring, because measuring and identifying TD was considered too difficult. Some of the teams (A, B, F, and G) used some basic information in the JIRA management tool to monitor how many issues had been assigned, and drew conclusions on the basis of that data. However, some teams (C, D, and F) used the SonarQube tool for TD monitoring.

We did not observe any specific calculations for TD prioritization, as prioritization was mostly based on hunches and previous experience and knowledge regarding the code base. The things that were taken into consideration when making a decision about TD prioritization were often based on scalability, business value, use of a feature, and customer effect, but they did not contain any exact numerical values.

TD communication was in a good shape in most of the cases (A, B, C, D, E, and H). The management and development team had a good TD communication structure, where the team manager had sufficient technical knowledge. However, we also saw cases (E, G), where the development team felt that the current communication about TD issues was lacking a lot. The developers felt that all the development time went to new features, and there was no time allotted by the business people to conduct refactoring of old solutions.

Almost every studied development team had set up coding standards to prevent TD. However, they were not always followed in reality, as they had been labeled as recommendations. Every development team also tried to catch bad design and solutions by implementing a code review practice to ensure the quality of the developed code before it would go to production. However, this was not always possible, because the review process was time-consuming, and effort had to be assigned to new features that were more important to the development team.

TD representation/documentation was done in three different ways: a development team with a unique TD backlog (Case E), development teams with quality/development backlogs consisting of TD issues (Cases C, D, and H), and development teams not using any backlog for TD items (Cases A, B, F, and G).

TECHNICAL DEBT MANAGEMENT FRAMEWORK

We developed a TDM framework based on the analysis of the eight studied development teams. The framework is presented in Table 5. The framework explains the activities, practices/tools, stakeholders, and responsibilities of TDM. After analyzing individual cases, we started to compare the cases to understand the similarities and differences of approaches and practices in TDM activities. We took all the approaches and practices found in the analysis and put them into the same table (Table 4) to understand how each activity was conducted in general, across the cases. We observed that all practices had a defined responsibility. We were therefore able to add also the responsible person to each TDM activity. When we had identified all the TDM approaches, practices and responsibilities, we started to compare the cases. During this comparison we realized that there was a lot of variation in the TDM approaches and practices.

Table 5: TDM framework.

TDM	TD repayment	TD prevention	TD	TD identification	TD measurement	TD monitoring	TD	TD prioritization
activity/TDM			representation/				communication	
levels			documentation					
Organized (Level 3)	Continuous repayment with monthly assigned percentage of the development tasks.	Mandatory prevention practices used by the team. Continuous practice during development.	Documentation is a mandatory practice in development. Issues are documented in a separate TD backlog.	Continuous identification conducted manually and/or with tools during development	Continuous measurement during development. Data analysis (various data used (e.g. quality, performance)). Assisted with tools.	Continuous monitoring during development with various data (e.g. quality, performance). Tools used to support.	Continuous discussions/ meetings about TD issues with all the necessary stakeholders involved.	Prioritization conducted continuously during development. Prioritization follows a specific method or model.
Received (Level 2)	Repayment during normal development tasks and previously identified repayment tasks. Repayment conducted based on current needs.	Optional prevention practices. Not mandatory to use, but recommended. Conducted based on current time constraints.	Documentation an optional practice, but recommended. Issues documented in a general development backlog without TD id.	Identification optional during normal development. Conducted based on current time constraints.	Measurement an optional practice. Measurement done with simple data (number of TD issues) from development, and the data not necessarily used for other activities.	Monitoring based on simple data (number of TD issues). Conducted occasionally.	Discussions/ meetings organized only with some stakeholders.	Prioritization based on hunches and rough estimations based on previous experiences. Prioritization done in a simple way without any specific model.

Unorganized (Level 1)	Repayment not conducted at all or only when it is not possible to avoid the issue any longer.	Prevention not assigned as part of the development practices. Conducted only occasionally.	Documentation not part of development. Issues are left in developers' own minds and notes.	Identification practices not assigned as part of development. Conducted only when issues occur.	Measurement not part of development practices.	Monitoring not part of development practices.	TD not a topic in discussions/ meetings and often handled only in coffee table discussions.	Prioritization not conducted, and decisions done without reasoning or discussions.
Responsibility for activity	Development team, software architect(s)	Development team, software architect(s)	Development team, software architect(s)	Development team, software architect(s)	Software architect(s), team manager	Software architect(s), team manager	Development team, software architect(s), team manager	Software architect(s), team manager
Practices / tools for activity	Refactoring, redesigning, rewriting	Coding standards, code reviews, Definition of Done.	Technical debt backlog/list, Documentation practice, project management tool (JIRA, Wiki)	Time reservation for manual code inspection. Use of code analysis tools (SonarQube, CheckStyle, FindBugs).	Data from measurement tools (SonarQube) and data from project management tools (JIRA, Wiki).	Monitoring tools (SonarQube). Project management tools (JIRA, Wiki)	Specific TD meetings, TD included in discussion topics.	Cost/Benefit model, Issue rating

The results indicated differences in the maturity of TDM. By the term maturity we mean the ability of the development team in TDM activities. Firstly, we identified cases where a TDM activity was not at all conducted during the development. We defined this as the lowest level of maturity, where a development team does not conduct a particular TDM activity. Secondly, we identified development processes where TDM activities were organized and conducted continuously by the development teams as a part of their normal development process. We defined this as the highest level of maturity, where the TDM activity is an integral part of the continuous development process. These two extremes were identified as the lowest and highest levels of maturity. TDM activities that were conducted only sometimes and were not considered an important part of the continuous development process, were placed on a level between these two extremes. We used these maturities to assign every identified TDM activity with their own maturity levels. On the basis of the process described above, we developed a TDM framework divided into five sections: *TDM activities, TDM levels, TDM stakeholders, TDM responsibilities, and TDM approaches.*

We use the eight activities identified by Li et al. (2015a) as *TDM activities* in the framework. The TDM activities are *TD repayment, TD prevention, TD documentation, TD identification, TD measurement, TD monitoring, TD communication, and TD prioritization.* Based on our findings in the studied development teams, we believe that the eight TDM activities are suitable for giving an overall view on TDM. During the analysis of the cases, we identified some level of approach in each TDM activity. In addition, we were not able to identify any new TDM activities during the analysis of the cases.

The analysis revealed that the TDM activities were conducted at different maturity levels. For example, we observed that while one development team focused on and put effort to measurement and monitoring activities, another development team did not put any effort to them. We defined three *TDM maturity levels: unorganized, received,* and *organized.* A TDM activity can be considered *unorganized* when a software team does not put any effort to the activity or when the focus is minimal. A TDM activity can be considered *received* when the software team has acknowledgd the need for a certain TDM activity and when it already conducts it on some level. However, the activity is not yet considered as a constant one and only a few people conduct it occasionally. A TDM activity can be considered as *organized* when the development team has recognized the TDM activity as an essential part of software development, and it is conducted continuously by the whole development team.

We identified three main stakeholders and one additional stakeholder related to TDM. These stakeholders came from the responsibilities found in the cross-case analysis. The first stakeholder is the *development team,* which is responsible for software development. The development team is often responsible for the TDM activities that take place during the actual development of software. These activities are TD repayment, TD prevention, TD documentation and TD identification. The development team works with the code base, and is able to identify and refactor possible issues in the software. The development team is also responsible for TD prevention in terms of following coding standards and code review practices.

The second stakeholder is the *software architect*, who is responsible for the architecture of the software. The software architects have responsibilities in all TDM activities. The cross-case analysis revealed that software architects often acted as a central mind in TDM. This was because software architects often have the best overall view on the software and its design issues. Therefore, all TDM activities should be within the responsibility of software architects.

The third stakeholder is the *team manager*, who is responsible for managing the development. We observed that the team manager was mainly responsible for four TDM activities: TD prioritization, TD communication, TD monitoring, and TD measurement. The team manager did not often deal with activities that were directly related to technical development, and therefore his/her responsibility was only on the management activities that required data collection as well. The team manager has a lot of communication with the business stakeholders to understand to what direction the software is evolving. Therefore, the manager is highly involved in the communication about TD, when there is a need to change the software to a certain direction. This also has an effect on the prioritization of TD, because business changes need to be evaluated with TD issues, to understand what kind of development efforts the software will need in the future.

An additional stakeholder is the *business stakeholder* who communicates about the software needs to the team manager. Business stakeholders are not necessarily directly related to TDM, but the needs coming from the business stakeholders do have an effect on the TDM activities. The business need e.g. for a new feature may change the current TD repayment activity or TD prioritization.

We also identified various approaches for each TDM activity. The approaches varied from practices conducted by the whole development team to practices conducted by a single person. We also made observations about the tools used to support the TDM approaches. The practices, models, methods and tools are presented in the framework in the section *approaches to activity.*

The framework can be used by software development companies to improve and evaluate internal and external processes regarding TDM. However, we cannot claim that working on the highest level of the TDM framework will reduce TD or produce healthier software. It is possible that a development team conducting refactoring only when necessary has less TD in their software than a development team that conducts all TDM activities continuously. Instead, we believe that using the framework will increase the visibility and knowledgeability regarding TD in the software, which can be used for smarter and safer decisions in TD reduction and management.

It is also important to mention that this framework is presented only at a high level, and it has been derived from the eight studied software development teams. Therefore, other researchers should improve this framework by adding approaches, responsibilities, levels, and activities that were not included in this study.

DISCUSSION

RQ1.1-1.2: What TDM activities are used in the studied development teams? What methods, models, practices or tools do the studied development teams' use for each TDM activity?

Commonly used activities

Communication

The most usual TDM activity in the studied development teams was *communication.* TD was an important discussion topic in most of the development teams. This is not a surprise, considering the popularity of TD research in the past few years (Li et al., 2015a). The biggest issue with TD communication has been the gap between technical and non-technical stakeholders (Klinger et al., 2011). Communication related to TD issues does not often transfer from the development team to the business stakeholders, which leads to TD issues not receiving the required time to get fixed (Yli-Huumo et al., 2014). Our observations also support the fact that the starting point for successful TDM is good TD communication. If the development team does not have any communication of TD, it is difficult to gain any benefit from the other TDM activities. Most of the studied development teams had organized TD communication successfully, which also helped in the other TDM activities. Simply taking TD as a topic in various meetings and discussions between the stakeholders can already improve TD communication. Especially a product manager with high business and technical competence can work effectively as a middle-man between business stakeholders and development teams, and improve communication related to TD.

Occasionally used activities

Repayment

TD repayment was conducted with refactoring, rewriting, and redesigning practices in the studied development teams. Similar practices identified in a study by Codabux and Williams (2013) were reengineering and repackaging. Even though all the practices mentioned for TD repayment had a similar goal of improving the solutions in the code base, it is still important to understand that they were not the same practices. Refactoring, which is a known concept in the literature and probably most commonly used technique for code improvement, can be described as a practice to improve code structures without changing the existing functional behavior of a program (Fowler et al., 1999). Redesigning can be an act to change the solution for example with a better and faster algorithm, while rewriting is an act to re-implement a large portion of an existing solution without re-using the previous source code. It is important to understand the differences between the concepts. Using *refactoring* as a term to describe large-size *rewriting* of a software feature can be misleading for some stakeholders in the development. Understanding the differences between TD repayment practices can improve especially TD communication, when all stakeholders understand the nature of the required improvements and the resources needed.

There are many strategies for conducting TD repayment. A development team can either choose to repay TD continuously, occasionally, or not at all. The decision to choose the repayment strategy emerges from the question "**do we have technical debt?**" In a case where the development team is fighting with a large TD, it would be wise to have a systematic way to repay TD back continuously to avoid a crisis in the future. In a case where the development team has only little known TD, it is possible to repay TD occasionally e.g. during normal development. Development teams can also choose not to repay any known TD, if they do not see any good reason for it.

Companies react differently to TD repayment. Some teams opt to reduce TD by a certain percentage every month, while some teams opt to focus on new features, and leave TD reduction to minimum (Power, 2013). Our observations in TD repayment strategies suggest that there is not necessarily one right TD repayment strategy and practices. The decision for the strategy has to be made on the basis of the current needs and understanding of the significance of TD in the software product.

Prevention

TD prevention activities happened only occasionally during development. Practices used for TD prevention included coding standards, code reviews, and the Definition of Done. A set of other practices for TD prevention have been identified in other studies (Codabux et al., 2014; Krishna and Basu, 2012). These practices include approaches such as *education and training, pair programming, test-driven development, refactoring, continuous integration, conformance to process and standards, tools, and customer feedback* (Codabux et al., 2014). Code reviews, where another developer checks your code can be used to prevent bad solutions from getting to the code base (Baker, 1997; Kemerer and Paulk, 2009), while setting up coding standards/guidelines for the development team to ensure as much cohesion as possible during the development (Green and Ledgard, 2011) can improve understandability and learnability.

TD prevention can be seen as one of the most influential activities of the eight TDM activities that a development team can conduct. When the development team has set up mandatory coding standards, assisted with e.g. code reviews and Definition of Done practice, it is possible that the amount of TD that gets to the code base will decrease (Davis, 2013). When TD is prevented as much as possible, it also helps other TDM activities. In addition, setting up TD prevention practices helps especially in catching unexperienced developers' 'not-so-good' solutions.

Even though the benefits of TD prevention are quite clear and simple to implement in real-life software development, we observed that they are still not necessarily used. The biggest issue was that they were conducted only occasionally, because they are not mandatory. The software development teams in this study mentioned having coding guidelines and reviews set up, but they were not often used. There are possible reasons for the development teams not using TD prevention practices. First, working with strict standards and guidelines in software development can sometimes be exhausting and annoying for developers, when they are not allowed to use their own creativity in the development, but must follow strict guidelines instead. Second, adopting TD prevention practices requires resources. Using various TD prevention practices requires time and competence, which are always taken away from something else.

Representation/documentation

TD representation/documen- tation was conducted only occasionally. There can be several reasons for why developers do not conduct documentation. In tight schedules documentation is often not seen as a useful practice, and therefore writing TD documentation can be seen as waste of time. Developers may also value documentation differently, and they document only issues that they personally think are important (Lethbridge et al., 2003). The biggest reason why TD representation/documentation was lacking in our cases was that TD was not generally considered as something that could/should be documented.

The development teams had a variety of approaches for documenting TD. Some teams had a specific TD list, which consisted of TD issues only and nothing else. Some teams used a normal development backlog as the place to store TD issues. The tools used for these two approaches were JIRA and Wiki, which made the data available for everyone. There were also teams that did not use any documentation for TD issues, and just decided to leave them as common knowledge in the development team.

We believe that TD representation/documentation is essential for a successful TDM strategy. When TD issues are not stored, it is highly possible that they will be forgotten at some point and will never be repaid. Without proper tracking and documentation of architectural changes and issues, it is also extremely challenging to quantify TD (Klinger et al., 2011). The inability to quantify TD also creates more challenges to other TDM activities, such as *communication, repayment, monitoring, and measurement,* due to the lack of TD data.

Documentation is a valuable practice that improves understandability and communication (Das et al., 2007; Forward and Lethbridge, 2002). Therefore, adopting even a simple documentation practice for TD representation/documentation improves other TDM activities and the overall TDM strategy. A systematic process to document and store all the TD issues can be used for creating a systematic TD repayment strategy (Lim et al., 2012).

Identification

TD identification was conducted occasionally during the development. In manual identification a person tries to locate the sources of a TD problem. Also tools can be used to find bad code. Most of the identification in the studied cases was conducted manually because of lacking tools or knowledge about them. Some development teams used tools like SonarQube, CheckStyles, and FindBugs to scan the code base to find possible complexities and badly developed code.

TD can be completely different for different development teams. Some development teams consider smaller issues, such as bugs or single line errors, to be TD. These types of smaller errors are simpler and easier to fix and they can be found with tools developed to scan the source code, such as SonarQube. Identifying issues found with these tools can mean for some development teams that TD has been identified, and they will use this TD data for other TDM activities.

However, the challenge in identification is that TD is not just related to simple errors, but especially to the architectural and design issues of software. It is challenging to identify this type of TD with tools. The challenge is how the tools tackle architectural or structural issues and technology gaps (Kruchten et al., 2012a). This was also mentioned by the architects and developers who did not have any tool available to find the types of issues that required manual identification. This issue has also been raised in a previous TD study (Zazworka et al., 2014), questioning how TD issues could be identified from the code base. It seems that TD identification is often done during the actual development, where a developer notices that something bigger might be wrong in some part of the code base. An interesting question related to TD identification is whether developer-identified TD should be considered as "real TD", while tool-identified TD should not, because it is not necessarily related to the effects of external (such as customer and market) changes in the software architecture (Zazworka et al., 2013).

Prioritization

Another occasionally used activity for the development teams was TD prioritization. When TD issues were identified, there was no precise model or method used to calculate or estimate the effects or costs of the TD. The literature has suggested approaches for TD prioritization (Eisenberg, 2012; Seaman et al., 2012; Theodoropoulos et al., 2011; Zazworka et al., 2011a). Some of the approaches are based on calculating technical values (e.g. duplicate code, test coverage, rules compliance, code comments etc.), some take aspects from the finance environment, such as cost-benefit analysis into consideration, while some use software quality attributes for the evaluation.

In our cases, the estimation and prioritization was just based on a hunch and previous knowledge of the person. The reason was that calculating technical things like scalability and further maintenance is extremely difficult, as business items like plans and business value have to be considered as well. Therefore, the prioritizations were often assessed on a low/medium/high scale or using story points to estimate the importance and effort of TD, based on hunches and rough estimations.

Ramasubbu et al. (2015) describe TD prioritization with three dimensions: *customer satisfaction needs, reliability demanded by the business, and probability of technology disruption.* These dimensions are essential for decisions, but quantifying these with exact numbers is extremely difficult. Prioritization can also be based on customer needs, but this can leave the most important TD from the technical perspective out of sight (Codabux and Williams, 2013). These prioritization issues exist also in requirements prioritization (Lehtola and Kauppinen, 2006).

Rarely used activities

Measurement

TD was measured rarely in the studied cases. The only identified measurement practices used either data available in project management tools (JIRA, Wiki), or a specific tool to measure TD (SonarQube). The data gathered from JIRA consisted usually of simple data only (reported TD issues, number of bugs etc.), which was used to get some level of understanding about the status of TD. The usefulness of this data could be questionable. For example a decrease of TD issues from 50 to 48 in one month does not necessarily mean that TD has been reduced, because there may exist unidentified TD issues. Some development teams used also e.g. quality and productivity as a measurement to see in which direction the software was going.

The data gathered with tools (e.g. SonarQube) provides an estimate of TD based on calculations. This type of data could be easier to interpret in development and management. For example, SonarQube calculates TD from seven deadly sins (SonarQube, 2015), each one representing a major quality item: bad distribution of the complexity, duplications, lack of comments, coding rule violations, potential bugs, no unit tests or useless ones, and bad design (SonarQube, 2015). Some of the development teams in the studied cases used this value to get an estimate of TD, which was followed during the development.

An estimate based on a tool should be more accurate, faster and reliable compared to an estimate based on simple data. However, TD measurement has the same problem as TD identification: "*what technical debt do you want to estimate*?" When a development team considers for example the criteria in SonarQube (2015) as TD, in can guide TD management and other TD activities. However, TD can also be considered to consist of issues of a larger scale, such as architectural or structural issues and technology gaps (Kruchten et al., 2012a). There are not necessarily any automatic tools available to measure these issues of a larger scale.

This can be currently seen as the biggest problem and challenge in TD measurement. There are no valid tools to measure larger TD issues related to the deep architectural structures of software. Therefore, most TD measurement is done on the basis of human evaluation, which can be seen as a challenge especially in decision-making.

Monitoring

Similar to TDM measurement, TD monitoring was also conducted rarely. The lack of TD monitoring is also related to the rare occurrences of TD measurement. Without any measurable TD data from the software, it is also almost impossible to monitor anything related to TD. Most of the TD monitoring was based on data derived from project management tools (JIRA, Wiki) or specific tools to measure TD (SonarQube). The team members responsible for monitoring TD used this data to monitor how TD was increasing or decreasing during the development, and used that information to assign work in other TDM activities.

TD monitoring and tracking is one of the most vital TDM activities (Ernst et al., 2015). Without monitoring, the development team is not able to have any reasoning for other TDM activities. One of the questions related to TD monitoring that can be seen as major obstacle is "what should you monitor?" Tools may help in estimating technical aspects, such as bad distribution of the complexity, duplications, lack of comments, coding rule violations, potential bugs, and lack of unit tests. However, an essential part of TD monitoring is also monitoring the overall quality of the software and the productivity of the development team. Evaluating how a large-scale architectural change affects the developers' productivity or the overall quality makes it possible to reason why some TD issues are important to repay or not.

RQ1.3: Are there any maturity differences on adopting TDM activities between development teams?

In some development teams TDM focused on only two to three activities, while some development teams conducted all eight TDM activities. Some development teams opted to use tools for the activities, while some teams did not have knowledge of available tools. Some development teams opted to conduct activities continuously, while some teams did it just occasionally.

The biggest maturity differences were in TDM activities that were conducted mostly by the development team (repayment, prevention, representation/documentation, and identification), while the least differences were in activities done mostly by the software architects and the team manager (measurement, monitoring, prioritization). We suggest that this was because the activities conducted mostly by the software architects and team managers were considered the most challenging, and there was not necessarily many known tools or practices available, which resulted in the fact that activities were not often conducted.

RQ1.4: What are the biggest challenges in TDM?

Lack of tools

One of the main challenges in TDM is the lack of tools. TDM was mostly conducted as human activity, instead of using automated or enabling tools. Ernst et al. (2015, p.?) state that developers "*desire standard practices and tools to manage technical debt that do not currently exist*". If most of the current TDM activities are done with rough estimations and are based on hunches, instead of tools and models based on precise data from specific tools, there is a risk that the choices made for TD reduction and management are not always the most optimal ones. In addition, conducting TDM activities without tools is time-consuming, and the addition of tools would provide faster TDM activities.

As Ernst et al. (2015, p.?) comment, "*tooling is a necessary component of any technical debt management strategy*", we also believe that an important research area currently in TDM is the research and development done for tools designed to tackle different TDM activities. The development of new tools especially for identification, measurement and monitoring activities can and should be beneficial and should be in a high priority in future research related to TDM.

Knowledge of TD priorities

Unlike the challenge with tools, TD prioritization is not necessarily as much dependent on tools, even though TD prioritization needs data input from other TDM activities to support the decision-making. However, one of the current main challenges of TDM is TD prioritization. The challenge is the lack of models and methods to prioritize TD issues successfully. There are no proper solutions to understand and explain why some TD items should be a priority to the development team over other TD items. Some type of technical debt can be important for a development team to fix, while a similar type of technical debt is not seen as a problem for another team. Some papers (Eisenberg, 2012; Seaman et al., 2012; Theodoropoulos et al., 2011; Zazworka et al., 2011a) discuss how TD issues should be prioritized on various levels. They include ideas and suggestions of how to prioritize TD issues, but they have not been thoroughly tested empirically, or they do not take all the aspects related to TD prioritization into consideration, including both technical and business needs. In prioritization both technical and business needs need to be covered.

We believe that TD prioritization as an activity is currently lacking models and methods that take both the technical and business needs of TD into consideration. The development teams in our study had a hard time prioritizing TD issues, because they had no model or method for doing it properly. Therefore, prioritization was mainly done just by the opinions of single persons, based on hunches and previous experiences, instead of estimations and measurements based on some precise data. There are cases where developers may have an idea of how to improve some part of the architecture to decrease complexity or increase velocity. However, if this improvement in architecture does not bring any value for the customers, it may not be prioritized as high as it should be from the technical perspective. On the other hand, a minor TD issue with lots of work and a high value to a customer could be prioritized high, since it has business value. This is a current challenge in TDM, because knowledge about the most important TD issues to fix may be missing, which may result in wrong decisions. The development of new models or methods for TD prioritization would help development teams to explain to the business people the real benefits of technical improvements more clearly, based on exact values (e.g. time, quality, maintenance, productivity, business value).

Having a proper mindset with TDM

One of the challenges is the mindset of the developers. The goal of TDM is to provide practices and tools to manage and reduce TD during software development (Li et al., 2015a). This obviously requires more effort on the already existing practices of tracking down and fixing issues to make technical improvements. Conducting TDM takes time, and it will have an effect on other software development activities. Instead of designing and developing a new feature, it could more useful to identify a badly designed code manually. There is a possibility that some stakeholders see this as a waste of time. Therefore, the mindset towards TDM can sometimes be negative, and the developers or managers just want to focus on developing something new, which will lead to the use of hotfixes and quick solutions.

One of the challenges in TDM is to get the whole organization/team included in TDM with a proper mindset. Instead of only a few people documenting TD issues to the backlog or taking part in TD communication, it is important that every member of the team contributes to TDM. This way all the TDM activities will support each other successfully.

Time-consuming TDM

We also observed that TDM is time-consuming. Adopting new TD processes and tools can create more work on top of the existing development process. Therefore, it may difficult to justify the real need for TDM and its benefits. For example, why should the development team have mandatory coding reviews or documentation practices, if they take time away from other important development practices, and there is no guarantee that they would provide immediate benefits? In addition, conducting e.g. manual code inspection takes a lot of time, and its benefits are uncertain. Therefore, adopting activities that require more time and resources to be successful, can be hard to justify. This is the reason why there is an urgent need to provide more evidence of TDM. Doing research on the benefits of conducting code reviews, on how documentation helps in TD visibility, or how manual code inspection can offer a possibility to detect serious architectural issues, can bring justification for the reasons to have TDM, which will give confidence to the development teams to allocate more time and resources for TDM.

Limitations of the study and threats to validity

Generalization of the results

A case study does not provide statistical generalizability (Yin, 2003), i.e. a case study with a limited number of cases cannot be generalized over a population. We, however, consider generalization as *theoretical* (Lee and Baskerville, 2003), i.e. abstraction from concrete events and actions to theoretical constructs. Case studies are generalizable to theoretical proportions, not populations or universes. We believe that the theoretical implications of this study are needed for creating a more focused approach to TDM.

Construct validity

The threats to the validity of a case study can be divided to four aspects: construct validity, internal validity, external validity, and reliability (Runeson and Höst, 2008). Construct validity reflects 'to what extent the operational measures that are studied really represent what the researcher has in mind and what is investigated according to the research questions' (ibid., p. 153). To improve construct validity in this study, the data collection protocol was reviewed, discussed, and corrected if necessary by all the authors. During the interviews, we also put a lot of emphasis on the explanation of each research question, and tried to improve the fact that both the interviewer and interviewee had similar understanding of the research topic. In addition, most of the interviews were conducted by two authors. This increased the possibility for the other interviewer to correct possible misunderstandings during the interviews. We also let the interviewees review the first draft of the paper, in order to identify issues in construct validity.

One limitation of the study is the difference in the interview structure between the first and second round interviews. As the first round interviews were conducted roughly one year before the second round interviews, and the interview structure was changed between the rounds, the collected data was not congruent. The first round interviews were analyzed first with a different data coding protocol, but we reanalyzed them afterwards with the same data coding protocol as with the second round interviews, to ensure the same coding process.

Internal validity

Internal validity is a concern when causal relations are examined. The concern is being certain that when a causality between x and y is found, factor z is not included, which we did not identify during the interviews (Runeson and Höst, 2008). Improving internal validity in case studies is challenging, because it is sometimes hard to know if there is some underlying reason for the causalities. We used semi-structured interviews to gain more in-depth knowledge related to the data in the studied cases. Therefore, when we were not completely satisfied with the gained data, we could ask more specific questions to understand the factors related to the causalities better. In addition, we were also able to communicate with the interviewees after the interviews, if we had some smaller additional questions about issues related to the data analysis.

External validity

External validity is concerned with 'to what extent it is possible to generalize the findings, and to what extent the findings are of interest to other people outside the investigated case' (Runeson and Höst, 2008, p. 154). One limitation of this study was the number of the studied software development teams and the fact that all of them were from the same organization. Obviously, adding more software development teams from several other organizations, the theory and framework could be possibly extended by adding new data. The goal of this study was not to create a complete and generalizable framework for TDM. Instead, the goal was to understand how the selected software development teams were managing TD in their current development environment. Therefore, the developed framework is not necessarily generalizable, because the data was derived only from one organization. However, the framework can be used for future research, and it can be improved and extended by adding new data from other empirical sources.

Reliability

Reliability is concerned with 'to what extent the data and the analysis are dependent on the specific researchers' (Runeson and Höst, 2008, p. 154). One limitation of this study is the semi-structured interview approach. In the semi-structured approach, the interview questions are often open-ended. Therefore, the answers from different interviewees can vary a lot, and the discussion during the interviews can be different in each interview session. In a situation where another researcher conducts the study, the data from the interviews will not necessarily be exactly the same. However, we improved the reliability of the study by designing and describing the data collection, data coding, and data analysis process carefully, which makes it more repeatable to other researchers.

Implications for future research

On the basis of our findings we believe that TDM in software development has similarities to the characteristics of the capability maturity model (CMM) (Paulk et al., 1993). There are similar differences in the maturity of TDM across projects and companies. The CMM was originally developed to present a set of recommended practices to enhance software development and maintenance capability. The fundamental concepts of CMM are capability, performance and maturity. The five levels in CMM are initial (chaotic), repeatable, defined, quantitatively managed, and optimizing (Paulk et al., 1993). A similar maturity model to CMM is also adaptable in TDM, where development teams have different TDM maturities in activities and practices. This kind of maturity as a concept has been applied to other processes and domains as well (De Bruin et al., 2005).

It is important to point out that our results do not show if there are any advantages or disadvantages in using some specific approaches or their combination. The success of TDM is not necessarily related to the number of approaches that a development team uses. It is possible that development teams conducting refactoring only when it is necessary have a less TD than development teams that monitor and measure TD constantly. However, we believe that having defined and structured

TDM activities and approaches can increase the visibility and knowledge regarding TD in software and projects. Therefore, we see the development of the TDM maturity model beneficial for both practice and research. Future research could focus on identifying TDM maturity levels and developing a practice-oriented maturity model, to improve the visibility and manageability of TD in software projects.

CONCLUSION

This study explored how software development teams manage technical debt in a real-life environment. We used the exploratory case study method suggested by Runeson and Höst (2008) to study eight software development teams in one large organization. For the analysis of technical debt management, we used the eight activities identified by Li et al. (2015a). We interviewed 25 persons to identify the processes, techniques and tools used for technical debt management.

We found that technical debt management was conducted at various levels. Some of the teams did not have any clear strategy or tools to manage and reduce technical debt, while some teams had defined structured processes to reduce, monitor, measure, and manage their technical debt. We also observed that there exist several challenges of technical debt management, which software development teams have to understand and acknowledge.

The study produced a technical debt management framework that describes the management activities, stakeholders and responsibilities on three levels and approaches/practices/tools used in them. The framework can be used for the definition of activities included in TDM, and how the activities are divided between the stakeholders.

Technical debt management has many similarities with the capability maturity model (CMM). We believe that the developed framework can serve as the basic element for researchers and practitioners in the development and improvement of technical debt activities.

ACKNOWLEDGMENTS

The authors would like to thank the company and their employees for participating in this research. The research has been carried out in the

Digile Need 4 Speed program, and it has been partially funded by Tekes (the Finnish Funding Agency for Technology and Innovation).

Appendix A

1. General information

1.1 Respondent's name:

1.2 Email:

1.3 Role in company:

1.4 Responsibilities:

1.5 Company name:

1.6 Organizational unit:

1.7 Industry sector:

1.8 Number of employees:

2. Technical debt

2.1 Have you heard of the term technical debt before?

2.2 Have you experienced situations where you had to take shortcuts in your projects, for example writing a code of lower quality or skipping a run of test cases to meet deadlines, and decided to fix them later?

2.3 Describe examples of shortcuts (technical debt) in your projects.

2.3.1 What kind of effect did they have right after?

2.3.2 How did they evolve during the software life cycle?

-Poor customer responsiveness?

-Long delivery times?

-Late deliveries?

-Lots of defects?

-Rising development costs?

-Frustrated and poor performing teams (bad productivity)?

2.3.2 What were the main reasons for you having to take these shortcuts?

2.3.3 Did you ever fix or make better the shortcuts you took?

2.3.4 Did you learn anything from these examples? Would you take the same shortcuts again? Why or why not?

2.4 Have you ever taken shortcuts in development because of pressure from business people or a customer due to deadlines?

2.5 Have you ever been "forced" to take shortcuts in a situation where business people did not necessarily understand the concept of technical debt and its effects on the project, and you thought it was a bad idea?

2.6 Are you willing to take shortcuts in development that will not cost much now but will cost more in the future, to meet the deadlines?

2.7 What is the business manager's opinion usually about taking these kinds of shortcuts?

2.8 How do you communicate between different organizational units about taking shortcuts in a project? Do you communicate about it with the customer?

2.9 How do you make decisions regarding taking shortcuts on projects?

2.10 Do you have any strategies as regards managing or reducing these shortcuts?

2.11 Do you think business people should include these kinds of shortcuts in their business strategy and budget?

2.12 How do you ensure that the quality level of your code is high and easily changeable to maintain?

2.13 How often do you do refactoring? Do you inform about it to the business people? What is their reaction to it?

2.14 What do you think are the positives and negatives of taking shortcuts?

2.15 Do you have any ideas on how your company (or companies in general) should take care of managing, finding, reducing and paying shortcuts?

2.16 Do you think technical debt actually exists? Is it a serious threat to software companies and should they pay more attention to it?

2.17 Do you think all shortcuts are bad and must be paid for at some point? How would you describe the difference between a good and a bad shortcut?

2.18 What software development methods or models do you use?

2.19 Have you used any other methods?

2.20 Do you think that there are differences between methods as regards taking shortcuts? Is it easier to manage with one or the other?

2.21 Do you have any other thoughts, comments, suggestions of what you have learned about technical debt / taking shortcuts in development what you would like to share?

Appendix B.

Appendix B Interviewee introduction

- Respondent's name:
- Respondent's name:
- Respondent's name:
- Email:
- Role in company:
- Responsibilities:

Introduction to the case

- Case history
  - What is the history of this team/case?
- Product history

- o What is the history of the product?
- o What has changed during the history?

Stakeholders of the case

- Describe what teams are included in this case (development, management etc.)?
- What are the sizes of the teams?
- Are there any problems with technical debt?
- Your team was interested in studying the topic of technical debt, so do you have some kind of a problem currently with technical debt?

Reasons for technical debt

- oIntentional technical debt: Strategic decisions to incur technical debt during a project.
- Do you have any examples of intentional technical debt in this case?
- Why was the intentional technical debt taken?
- oUnintentional technical debt: Lack of practices to retain the code quality level
- o Do you have any examples of unintentional technical debt in this case?
- Software development methodologies/processes/tools
- o What software development methodologies are you using (waterfall vs. agile)?
- o Do you think that software development methodology has any effect on technical debt?

Effects of technical debt

- How does technical debt affect you?
- Time-to-market?

- Lack of productivity?
- Lack of quality?
- Extra work?
- Bugs/Errors/Defects?

Management of technical debt

- TD repayment
- o How are you repaying technical debt back?
- o How has refactoring been organized in your team?
- o Do you refactor only when it is necessary or do you have a plan for it?
- TD identification
    - How do you identify technical debt?
    - Do you have any tool for it or do you do it manually?
- TD measurement
    - How do you measure technical debt?
    - Do you have any tool for it or do you calculate it manually from somewhere?
- TD monitoring
    - How do you monitor technical debt?
    - Do you have any tool for it?
- TD prioritization
    - How do you prioritize technical debts?
    - Do you do it based on a hunch and experience, or do you have a model/method for it?
- TD communication

    - How have you organized communication about technical debt?
    - Do you discuss technical debt often with the whole team?
- TD prevention
    - How do you prevent technical debt?
    - Coding standards?
    - Code reviews?
    - Definition of done?
- TD representation/documentation
    - Do you document technical debt issues in any way?
    - Do you have a separate technical debt backlog?

Improvements for current technical debt

- Possible suggestions for improvements
    - How would you like to improve you current practices regarding technical debt management?
    - Is there anything else you would like to say?

REFERENCES

1. Allman, E., 2012. Managing technical debt. Commun. ACM 55, 50–55. doi: 10.1145/ 2160718.2160733 .
2. Al Mamun, M.A., Berger, C., Hansson, J., 2014. Explicating, understanding, and man- aging technical debt from self-driving miniature car projects. In: 2014 Sixth In- ternational Workshop on Managing Technical Debt (MTD), pp. 11–18. doi: 10. 1109/ MTD.2014.15.
3. Alves, N.S.R., Ribeiro, L.F., Caires, V., Mendes, T.S., Spinola, R.O., 2014. Towards an ontology of terms on technical debt. In: 2014 Sixth International Workshop on Managing Technical Debt (MTD), pp. 1–7. doi: 10.1109/MTD.2014.9 .

4. Baker Jr., R.A., 1997. Code reviews enhance software quality. In: Proceedings of the 19th International Conference on Software Engineering, ICSE '97. New York, NY, USA. ACM, pp. 570–571. doi: 10.1145/253228.253461 .
5. Barney, S., Aurum, A., Wohlin, C., 2008. A product management challenge: creating software product value through requirements selection. J. Syst. Archit. 54, 576– 593. doi: 10.1016/j.sysarc.20 07.12.0 04 .
6. Boehm, B.W. , 2006. Value-based software engineering: seven key elements and ethical considerations. In: Biffl, S., Aurum, A., Boehm, B., Erdogmus, H., Grün- bacher, P. (Eds.), Value-Based Software Engineering. Springer, Berlin Heidelberg, pp. 109–132 .
7. Brown, N., Cai, Y., Guo, Y., Kazman, R., Kim, M., Kruchten, P., Lim, E., MacCormack, A., Nord, R., Ozkaya, I., Sangwan, R., Seaman, C., Sullivan, K., Zazworka, N., 2010. Managing technical debt in software-reliant systems. In: Proceedings of the FSE/SDP Workshop on Future of Software Engineering Research, FoSER '10. New York, NY, USA. ACM, pp. 47–52. doi: 10.1145/1882362.1882373 .
8. Charmaz, K. , 2014. Constructing Grounded Theory, second ed. SAGE Publications Ltd, Thousand Oaks, CA .
9. Codabux, Z., Williams, B., 2013. Managing technical debt: an industrial case study. In: 2013 4th International Workshop on Managing Technical Debt (MTD), pp. 8– 15. doi: 10.1109/MTD.2013.6608672 .
10. Codabux, Z. , Williams, B. , Niu, N. , 2014. A quality assurance approach to technical debt. In: Proc. Int. Conf. Softw. Eng. Res. Pract. SERP Steer. Comm. World Congr. Comput. Sci. Comput. Eng. Appl. Comput. WorldComp .
11. Cunningham, W. , 1992. The wycash portfolio management system (experience re- port). OOPSLA .
12. Das, S., Lutters, W.G., Seaman, C.B., 2007. Understanding documentation value in software maintenance. In: Proceedings of the 2007 Symposium on Computer Human Interaction for the Management of Information Technology, CHIMIT '07. New York, NY, USA. ACM doi: 10.1145/1234772.1234790 .
13. Davis, N., 2013. Driving quality improvement and reducing

technical debt with the definition of done. In: Agile Conference (AGILE), 2013, pp. 164–168. doi: 10.1109/ AGILE.2013.21 .
14. De Bruin, T. , Freeze, R. , Kaulkarni, U. , Rosemann, M. , 2005. Understanding the main phases of developing a maturity assessment model. In: Campbell, B., Under- wood, J., Bunker, D. (Eds.), Faculty of Science and Technology. Presented at the Australasian Conference on Information Systems (ACIS), Australasian Chapter of the Association for Information Systems, CD-ROM, pp. 8–19 .
15. Eisenberg, R.J., 2012. A threshold based approach to technical debt. SIGSOFT Softw. Eng. Notes 37, 1–6. doi: 10.1145/2108144.2108151 .
16. Ernst, N.A., Bellomo, S., Ozkaya, I., Nord, R.L., Gorton, I., 2015. Measure it? Man- age it? ignore It? software practitioners and technical debt. In: Proceedings of the 2015 10th Joint Meeting on Foundations of Software Engineering, ESEC/FSE 2015. New York, NY, USA. ACM, pp. 50–60. doi: 10.1145/2786805.2786848 .
17. Falessi, D., Kruchten, P., Nord, R.L., Ozkaya, I., 2014. Technical debt at the crossroads of research and practice: report on the fifth international workshop on manag- ing technical debt. SIGSOFT Softw. Eng. Notes 39, 31–33. doi: 10.1145/2579281. 2579311 .
18. Forward, A., Lethbridge, T.C., 2002. The relevance of software documentation, tools and technologies: a survey. In: Proceedings of the 2002 ACM Symposium on Document Engineering, DocEng '02. New York, NY, USA. ACM, pp. 26–33. doi: 10. 1145/585058.585065 .
19. Fowler, M., 2009. TechnicalDebtQuadrant [WWW Document]. URL http://martinfowler.com/bliki/TechnicalDebtQuadrant.html (accessed
20. 7.7.14).
21. Fowler, M. , Beck, K. , Brant, J. , Opdyke, W. , Roberts, D. , 1999. Refactoring: Improving the Design of Existing Code, first ed. Addison-Wesley Professional, Reading, MA .
22. Green, R., Ledgard, H., 2011. Coding guidelines: finding the art in the science. Com- mun. ACM 54, 57–63. doi: 10.1145/2043174.2043191 .
23. Griffith, I., Reimanis, D., Izurieta, C., Codabux, Z., Deo, A.,

Williams, B., 2014. The correspondence between software quality models and technical debt estimation approaches. In: 2014 Sixth International Workshop on Managing Technical Debt (MTD), pp. 19–26. doi: 10.1109/MTD.2014.13 .

24. Guo, Y., Seaman, C., 2011. A portfolio approach to technical debt management. In: Proceedings of the 2nd Workshop on Managing Technical Debt, MTD '11. New York, NY, USA. ACM, pp. 31–34. doi: 10.1145/1985362.1985370 .
25. Guo, Y., Seaman, C., Gomes, R., Cavalcanti, A., Tonin, G., da Silva, F.Q.B., San- tos, A.L.M., Siebra, C., 2011. Tracking technical debt #x2014; an exploratory case study. In: 2011 27th IEEE International Conference on Software Maintenance (ICSM), pp. 528–531. doi: 10.1109/ICSM.2011.6080824 .
26. Holvitie, J. , Leppänen, V. , 2013. DebtFlag: technical debt management with a devel- opment environment integrated tool. In: Proceedings of the 4th International Workshop on Managing Technical Debt, MTD '13. Piscataway, NJ, USA. IEEE Press, pp. 20–27 .
27. Kemerer, C.F., Paulk, M.C., 2009. The impact of design and code reviews on software quality: an empirical study based on PSP data. IEEE Trans. Softw. Eng. 35, 534 – 550. doi: 10.1109/TSE.2009.27 .
28. Klinger, T., Tarr, P., Wagstrom, P., Williams, C., 2011. An enterprise perspective on technical debt. In: Proceedings of the 2nd Workshop on Managing Techni- cal Debt, MTD '11. New York, NY, USA. ACM, pp. 35–38. doi: 10.1145/1985362. 1985371 .
29. Krishna, V., Basu, A., 2012. Minimizing Technical Debt: developer's viewpoint. In: International Conference on Software Engineering and Mobile Application Mod- elling and Development (ICSEMA 2012), pp. 1–5. doi: 10.1049/ic.2012.0147 .
30. Kruchten, P., Nord, R.L., Ozkaya, I., 2012. Technical debt: from metaphor to theory and practice. IEEE Softw. 29, 18–21. doi: 10.1109/MS.2012.167 .
31. Kruchten, P., Nord, R.L., Ozkaya, I., Visser, J., 2012. Technical debt in software devel- opment: from metaphor to theory report on the third international workshop on managing technical debt. SIGSOFT Softw. Eng. Notes 37, 36–38. doi: 10.1145/ 2347696.2347698 .

32. Lee, A.S., Baskerville, R.L., 2003. Generalizing generalizability in information systems research. Inf. Syst. Res. 14, 221–243. doi: 10.1287/isre.14.3.221.16560 .

33. Lehtola, L., Kauppinen, M., 2006. Suitability of requirements prioritization methods for market-driven software product development. Softw. Process Improv. Pract. 11, 7–19. doi: 10.1002/spip.249 .

34. Lethbridge, T.C., Singer, J., Forward, A., 2003. How software engineers use documen- tation: the state of the practice. IEEE Softw. 20, 35–39. doi: 10.1109/MS.2003. 1241364 .

35. Letouzey, J.-L. , 2012. The SQALE method for evaluating technical debt. In: Proceed- ings of the Third International Workshop on Managing Technical Debt, MTD '12. Piscataway, NJ, USA. IEEE Press, pp. 31–36 .

36. Letouzey, J., Ilkiewicz, M., 2012. Managing technical debt with the SQALE method. IEEE Softw. 29, 44–51. doi: 10.1109/MS.2012.129 .

37. Lim, E., Taksande, N., Seaman, C., 2012. A balancing act: what software practitioners have to say about technical debt. IEEE Softw. 29, 22–27. doi: 10.1109/MS.2012. 130 .

38. Li, Z., Avgeriou, P., Liang, P., 2015. A systematic mapping study on technical debt and its management. J. Syst. Softw. 101, 193–220. doi: 10.1016/j.jss.2014.12.027 .

39. Li, Z. , Liang, P. , Avgeriou, P. , 2015. Architectural technical debt identification based on architecture decisions and change scenarios. In: Proc. 12th Work. IEEEIFIP Conf. Softw. Archit. WICSA. .

40. Mantyla, M.V., Lassenius, C., 2009. What types of defects are really discovered in code reviews? IEEE Trans. Softw. Eng. 35, 430–448. doi: 10.1109/TSE.2008.71 .

41. McConnell, S., 2007. Technical Debt-10x Software Development | Construx [WWW Document]. URL http://www.construx.com/10x _ Software _ Development/ Technical _ Debt/ . (accessed 3.25.14)

42. Nord, R.L., Ozkaya, I., Kruchten, P., Gonzalez-Rojas, M., 2012. In search of a metric for managing architectural technical debt. In: 2012 Joint Working IEEE/IFIP Confer- ence on Software Architecture (WICSA) and European Conference on Software Architecture

(ECSA)., pp. 91–100. doi: 10.1109/WICSA-ECSA.212.17 .

43. Norton, M., 2009. Doc On Dev: Messy Code is not Technical Debt.
44. Paulk, M.C., Curtis, B., Chrissis, M.B., Weber, C.V., 1993. Capability maturity model, version 1.1. IEEE Softw. 10, 18–27. doi: 10.1109/52.219617 .
45. Power, K., 2013. Understanding the impact of technical debt on the capacity and velocity of teams and organizations: viewing team and organization capacity as a portfolio of real options. In: 2013 4th International Workshop on Managing Technical Debt (MTD), pp. 28–31. doi: 10.1109/MTD.2013.6608675 .
46. Ramasubbu, N., Kemerer, C.F., Woodard, C.J., 2015. Managing technical debt: insights from recent empirical evidence. IEEE Softw. 32, 22–25. doi: 10.1109/MS.2015.45 . Robson, C. , 2002. Real World Research, second ed. Wiley- Blackwell .
47. Runeson, P., Höst, M., 2008. Guidelines for conducting and reporting case study research in software engineering. Empir. Softw. Eng. 14, 131–164. doi: 10.1007/ s10664- 008- 9102- 8 .
48. Seaman, C.B., 1999. Qualitative methods in empirical studies of software engineer- ing. IEEE Trans. Softw. Eng. 25, 557–572. doi: 10.1109/32.799955 .
49. Seaman, C., Guo, Y., Zazworka, N., Shull, F., Izurieta, C., Cai, Y., Vetro, A., 2012. Using technical debt data in decision making: potential decision approaches. In: 2012 Third International Workshop on Managing Technical Debt (MTD), pp. 45–48. doi: 10.1109/MTD.2012.6225999 .
50. Seaman, C., Nord, R.L., Kruchten, P., Ozkaya, I., 2015. Technical debt: beyond defi- nition to understanding report on the sixth international workshop on manag- ing technical debt. SIGSOFT Softw. Eng. Notes 40, 32–34. doi: 10.1145/2735399. 2735419 .
51. SonarQube, 2015. http://www.sonarqube.org/evaluate-your-technical-debt-with- sonar/ Accessed: 2015-10-29.
52. Strauss, A. , Corbin, J.M. , 1998. Basics of Qualitative Research: Techniques and Proce- dures for Developing Grounded Theory. SAGE Publications .
53. Tamburri, D.A., Kruchten, P., Lago, P., Van Vliet, H., 2013. What

is social debt in software engineering? In: 2013 6th International Workshop on Cooperative and Hu- man Aspects of Software Engineering (CHASE), pp. 93–96. doi: 10.1109/CHASE. 2013.6614739 .

54. Theodoropoulos, T., Hofberg, M., Kern, D., 2011. Technical debt from the stakeholder perspective. In: Proceedings of the 2nd Workshop on Managing Technical Debt, MTD '11. New York, NY, USA. ACM, pp. 43–46. doi: 10.1145/1985362.1985373 .
55. Tom, E., Aurum, A., Vidgen, R., 2013. An exploration of technical debt. J. Syst. Softw. 86, 1498–1516. doi: 10.1016/j.jss.2012.12.052 .
56. Yin, R.K. , 2003. Case Study Research: Design and Methods. Sage Publications, Thou- sand Oaks, Calif .
57. Yli-Huumo, J. , Maglyas, A. , Smolander, K. , 2015. The benefits and consequences of workarounds in software development project. 6th International Conference on Software Business .
58. Yli-Huumo, J. , Maglyas, A. , Smolander, K. , 2014. The sources and approaches to man- agement of technical debt: a case study of two product lines in a middle-size finnish software company. In: Jedlitschka, A., Kuvaja, P., Kuhrmann, M., Män- nistö, T., Münch, J., Raatikainen, M. (Eds.), Product-Focused Software Process Improvement, Lecture Notes in Computer Science. Springer International Pub- lishing, pp. 93–107 .
59. Yli-Huumo, J. , Rissanen, T. , Maglyas, A. , Smolander, K. , Sainio, L.-M. , 2015. The rela- tionship between business model experimentation and technical debt. 6th In- ternational Conference on Software Business .
60. Zazworka, N., Seaman, C., Shull, F., 2011. Prioritizing design debt investment op- portunities. In: Proceedings of the 2nd Workshop on Managing Technical Debt, MTD '11. New York, NY, USA. ACM, pp. 39–42. doi: 10.1145/1985362.1985372 .
61. Zazworka, N., Shaw, M.A., Shull, F., Seaman, C., 2011. Investigating the impact of design debt on software quality. In: Proceedings of the 2nd Workshop on Man- aging Technical Debt, MTD '11. New York, NY, USA. ACM, pp. 17–23. doi: 10.1145/ 1985362.1985366 .
62. Zazworka, N., Spínola, R.O., Vetro', A., Shull, F., Seaman, C., 2013. A case study on effectively identifying technical debt. In:

Proceedings of the 17th International Conference on Evaluation and Assessment in Software Engineering, EASE '13. New York, NY, USA. ACM, pp. 42–47. doi: 10.1145/2460999.2461005 .

63. Zazworka, N., Vetro, A., Izurieta, C., Wong, S., Cai, Y., Seaman, C., Shull, F., 2014. Comparing four approaches for technical debt identification. Softw. Qual. J. 22, 403–426. doi: 10.1007/s11219-013- 9200- 8

CHAPTER 12

CHALLENGES AND SUCCESS FACTORS FOR LARGE-SCALE AGILE TRANSFORMATIONS: A SYSTEMATIC LITERATURE REVIEW

Kim Dikert [a], Maria Paasivaaraa,[b], Casper Lassenius [a,b]

[a] School of Science, Department of Computer Science and Engineering, Aalto University, Finland

[b] Massachusetts Institute of Technology, Sloan School of Management, United States

ABSTRACT

Agile methods have become an appealing alternative for companies striving to improve their performance, but the methods were originally designed for small and individual teams. This creates unique challenges when introducing agile at scale, when development teams must synchronize their activities, and there might be a need to interface with other organizational units. In this paper we present a systematic literature review on how agile methods and lean software development has been adopted at scale, focusing on reported challenges and success factors in the transformation. We conducted a systematic literature review of industrial large-scale agile transformations. Our keyword search found 1875 papers. We included 52 publications describing 42 industrial cases presenting the process of taking large-scale agile development into use. Almost 90% of the included papers were experience reports, indicating a lack of sound academic research on the topic. We identified 35 reported challenges grouped into nine categories, and 29 success factors, grouped into eleven categories. The most salient success factor categories were management support, choosing and customizing the agile model, training and coaching, and mindset and alignment.

Keywords: Agile software development, Organizational transformation, Large-scale agile, Adopting agile software development, Challenges, Success factors, Systematic literature review

INTRODUCTION

Agile methods were originally designed for use in small, single-team projects (Boehm and Turner, 2005). However, their shown and potential benefits have made them attractive also outside this context, particularly both for larger projects and in larger companies. This despite the fact that they are more difficult to implement in larger projects (Dybå and Dingsøyr, 2009). Compared to small projects, which are ideal for agile development, larger ones are characterized by the need for additional coordination. A particular problem in applying agile to larger projects is how to handle inter-team coordination. Large-scale agile involves additional concerns in interfacing with other organizational units, such as human resources, marketing and sales, and product management. In addition, large scale may cause users and other stakeholders to become distant from the development teams. Despite these known problems related to large-scale agile, there is an industry trend towards adopting agile methodologies in-the-large (VersionOne, Inc, 2016; Paasivaara et al., 2013, 2014; Dingsøyr and Moe, 2014).

The State of Agile Survey that Version One has been conducting annually since 2007, has recently asked a few questions related to large scale as well, e.g. on scaling methods used and tips for success with scaling agile. According to the latest survey (VersionOne, Inc, 2016), 62% of the almost 4000 respondents had more than a hundred people in their software organization and 43% of all the respondents worked in development organizations where more than half of the teams were agile. Of course, the sample of this study is limited to a selected subset of companies and countries (of the almost 4000 respondents to the latest survey 65% were from North America and 26% from Europe). However, this indicates that there seems to exist a large number of companies that have taken or are taking agile into use in large-scale settings (VersionOne, Inc, 2016).

While the research literature contains several experience reports and some case studies on large-scale agile adoption, a systematic overview and synthesis of this growing body of research is still missing. Freudenberg and Sharp (2010) asked the industrial practitioners at the XP2010 conference to create a backlog of topics they think should be studied. The practitioners voted "Agile and large projects" as the top burning research question. Moreover, among the top ten items three focused on distributed agile development, which is relevant especially for larger organizations as they are often geographically distributed. In two recent workshops on large-scale agile development organized in XP2013 and XP2014 conferences, adoption of agile methods was one of the highlighted themes needing more research (Dingsøyr and Moe, 2013, 2014).

While research on agile software development is accumulating and maturing, and has provided a basis for conducting systematic literature reviews (Dybå and Dingsøyr, 2008; Jalali and Wohlin, 2012; Senapathi and Srinivasan, 2013; Kaisti et al., 2013), the area of large-scale agile development has not yet been studied through secondary studies. In this paper we start filling in this gap by presenting a systematic literature review of large-scale agile transformations.

BACKGROUND

Agile software development

Agile software development is a set of iterative and incremental software engineering methods that are advocated based on an "agile philosophy" captured in the Agile Manifesto (Fowler and Highsmith, 2001). While mostly repackaging and re-branding previously well-known good software development practices, the agile movement can be considered as an alternative to so called traditional software development methods. Traditional methods focus on up-front planning and strict management of change, but agile methods were designed to accept and efficiently manage change (Highsmith and Cockburn, 2001; Cockburn and Highsmith, 2001).

Agile methods have been both criticized and advocated, and research has shown that accommodating change may be a factor in both success and failure (Boehm, 2002). It has been shown that agile methods have improved satisfaction of both customers and developers, but on the other hand there is evidence that agile methods may not be a good fit for large undertakings (Dybå and Dingsøyr, 2009). A proposed solution is that each organization seeks its own balance of agile and plan driven methods (Boehm, 2002).

Two of the most popular agile methods are Extreme Programming (XP) and Scrum (Hamed and Abushama, 2013). Scrum is a method focusing on the project management viewpoint of agile development (Schwaber and Beedle, 2002), prescribing timeboxing, continuous tracking of project progress, and customer centricity. The XP development method is a collection of practices for enabling efficient incremental development (Beck, 1999). In practice, many agile development implementations combine the two in some way (Fitzgerald et al., 2006).

Adopting large-scale agile

The difficulty of introducing agile methods increases with the organization size (Dybå and Dingsøyr, 2008). The difficulty is partly related to size bringing higher organizational inertia which slows down organizational change (Livermore, 2008). Agile development is not founded on the use of individual tools or practices, but rather on a holistic way of thinking. Adopting agile often requires change of the entire organizational culture (Misra et al., 2010).

One significant difference between small and large-scale adoptions is that larger organizations have more dependencies between projects and teams. This increases the need for formal documentation and thus reduces agility (Lindvall et al., 2004). In addition to inter-team coordination, development teams must interact with other organizational units, which are often non-agile in nature. For instance, human resources unit may demand individuals to have strictly specified roles in projects (Boehm and Turner, 2005), or a change control board may inhibit the use of continuous integration or refactoring (Lindvall et al., 2004). All units affected by the agile transformation need to be informed and consulted, and the agile process must be adjusted according to their needs (Lindvall et al., 2004; Cohn and Ford, 2003; Boehm and Turner, 2005).

Agile methods also affect management and business related functions. A key challenge is that management must move away from life-cycle models and towards iterative and feature centric models (Nerur et al., 2005), which requires a change of mindset. The focus must be shifted from long-term planning to shorter term project planning (Misra et al., 2010), as agile methods emphasize that planning is only meaningful for the near future (Cohn and Ford, 2003). However, the lack of planning can be a concern as business and customer relationships often build on long term roadmapping. Enabling operation with shorter term planning requires educating stakeholders and reviewing contracting practices (Boehm and Turner, 2005).

Definition of large-scale agile

A brief literature search (Dingsøyr et al., 2014) identified previous interpretations of what large-scale agile is. Size had been regarded in terms of size in persons or teams, project budget, code base size, and project duration. The examples of cases that were called "large-scale" included 40 people and 7 teams (Paasivaara et al., 2008), project cost of over 10 million GBP and a team size of over 50 people (Berger and Beynon-Davies, 2009), a code base size of over 5 million lines of code (Petersen and Wohlin, 2010), and a project time of 2 years with a project scope of 60–80 features (Bjarnason et al., 2011). Based on their findings Dingsøyr et al. (2014) ended up measuring large-scale by the number of collaborating and coordinating teams: they categorized as large-scale 2-9 collaborating teams and as very large-scale over ten collaborating teams.

We identified a number of additional studies discussing large-scale agile software development and their interpretations of large-scale. All of these referred to the number of people involved. In early work on agile,

Fowler considers the Crystal methodology to be suitable for up to 50 people (Fowler, 2000). The same number has been reported as seen by practitioners and researchers as the size of the largest organization suitable for agile (Williams and Cockburn, 2003). Other studies have referred to agile projects including up to 50 people as small (Koehnemann and Coats, 2009), and considered a development project large if it had a staff between 50 and 100 people, including all project personnel (Elshamy and Elssamadisy, 2006). The largest numbers were 300 people across 3 sites (Moore and Spens, 2008). Participants of the XP2014 large-scale agile workshop gave very varying definitions for large-scale agile development (Dingsøyr and Moe, 2014), showing that what is seen as large-scale depends very much on the context and the person defining it.

Based on these findings, we defined large-scale to denote *software development organizations with 50 or more people or at least six teams*. All people do not need to be developers, but must belong to the same software development organization developing a common product or project, and thus have a need to collaborate. For instance, Scrum masters and software architects are counted when assessing the organizational size. As some studies present the number of teams rather than the number of people, we correspondingly defined large-scale to denote development efforts involving at least six teams. Having six teams with an average size of six to seven people, plus a number of supporting staff, can reasonably be considered to form an organization of 50 people. In this definition, we include both companies that as a whole focus on software development, as well as the parts of larger (non-software focused) organizations that develop software, e.g. in-house software development units of large non-software corporations.

RESEARCH METHOD

Research questions

Systematic literature reviews are a means of identifying, evaluating and interpreting all available research relevant to a particular research question, or topic area, or phenomenon of interest. They are appropriate for summarizing existing research, for identifying gaps in the existing literature, as well as for providing background for positioning new research

(Kitchenham, 2007). In this paper, we present the results of a systematic literature review on the topic of large-scale agile transformations. The review is positioned in, and utilizes the literature of, the field of software engineering. In the review, we study the following research questions:

- RQ1: What challenges have been reported for large-scale agile transformations?
- RQ2: What success factors have been reported for large-scale agile transformations?

While large-scale agile transformations could provide many viewpoints to use as research questions, we chose these two questions because we believe they represent the viewpoints which are likely to provide actionable insights to practitioners as well as researchers. The research questions are not intended to present complements to each other, but rather two distinct viewpoints that can highlight different characteristics in the transformations.

Research process

The research process consisted of four main stages depicted in Fig. 1. The selection of primary studies was done in two stages, first using keyword-based database searches to identify potentially relevant sources, and then manually filtering the search result. The manual filtering process was executed independently by the two first authors. Data extraction was done by qualitative coding of the selected primary studies by the first author. Finally, the results were elicited by aggregating and analyzing the coding of the primary documents. The entire process was audited and mentored by the third author.

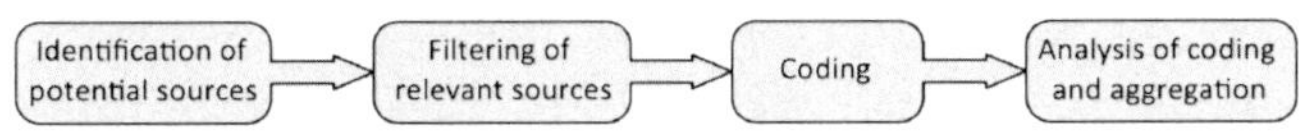

Figure 1: Outline of the research process.

Inclusion criteria

Based on the research questions and focus of the research, we defined four facets to guide our inclusion/exclusion decisions: agile software development, organizational transformation, large-scale, and empirical. Table 1 lists the facets and gives examples on matching topics and non-relevant topics. For a study to be included it needed to be relevant on all facets.

Table 1: Inclusion criteria.

Facet	**Relevant topics**	**Examples of non-relevant topics**
Agile software development	The organization develops software; the introduced method is agile	Agile manufacturing; Scrum in management boards
Organizational transformation	Presents insights about the transformation process	Comparison of before and after; how agile is currently used in large scale
Large scale	Agile practiced by a development organization of at least 50 people or 6 teams	Scaling up from small; a single agile team in a large setting
Empirical	Case studies, experience reports	Textbooks, student experiments, theory papers

Examples of topics excluded by the facet of *agile software development* are agile manufacturing and the application of Scrum outside software engineering, e.g., in management boards. In addition, we required that the organizational transformation was aimed at introducing agile methods, which excluded other development methods (Sagesser et al., 2013), and the use of agile methods in other contexts than software development (Hodgkins and Hohmann, 2007).

The facet of *organizational transformation* was interpreted so that the primary study had to present insights on the process of an agile transformation. Examples of excluded topics include comparing the original and agile development methods (Petersen and Wohlin, 2010), discussing the use of agile in a large enough organization but not describing how the new methods were introduced (Mishra and Mishra, 2011), and merely presenting agile tools in large-scale use (Kim and Ryoo, 2012).

Transformation and "scaling up" of agile practices in use are very closely related concepts, and in some cases they are one and the same. For instance, if transformation begins with a pilot and then is gradually rolled out in the organization, the process could well be seen as a "scaling up" journey. These kinds of journeys are included in our study. However, we excluded cases where an initially small agile organization grew organically (Maranzato et al., 2012), and discussions focusing on processes or tools without describing organizational change (Lyon and Evans, 2008).

The facet of *large scale* was interpreted as discussed in Section 2.3: a single software development organization with at least 50 persons or six teams. In some cases, the source presented very vague indicators on size. For instance, one case (Cloke, 2007) was included, as there were indications of large-scale considerations, although the size remained unclear. If there were no indications of size, the paper was excluded, e.g. (Miller and Haddad, 2012).

To complicate matters, some papers talked about "the team" in singular when referring to the organization (Hodgkins and Hohmann, 2007), making it nontrivial to judge whether the organization met the large scale criteria based on the choice of words of the author.

Further examples on exclusion by the *large scale* facet were cases with large organizations but only a single team adopting agile (Fulgham et al., 2011). We considered cases of single teams (although as part of a larger organization) unrelated to this research as our focus is on transformation of the entire (development) organization.

Also piloting cases that reported only single teams using agile, even though considering the whole organization would meet the large scale criteria, were excluded, e.g. (Scott et al., 2008). Finally, cases where the organization was growing to large scale, but did not meet the size criteria at the start of the transformation, were excluded.

According to the facet of *empirical* we excluded studies that did not discuss a distinguishable real world case. We excluded textbooks, studies merely presenting theories, as well as studies that did not include any case organization. Moreover, we excluded studies on the benefits or limitations of agile in general. Lastly, we excluded student experiments as it is implausible to draw conclusions on the organizational dynamics in our large-scale context based on experiments.

Table 2: Databases included in search, and number of matched articles.

Database	URL	# of matches
IEEExplore	http://ieeexplore.ieee.org	745
ACM	http://dl.acm.org	168
Scopus	http://www.scopus.com/home.url	1596
Web of knowledge	http://apps.webofknowledge.com	786

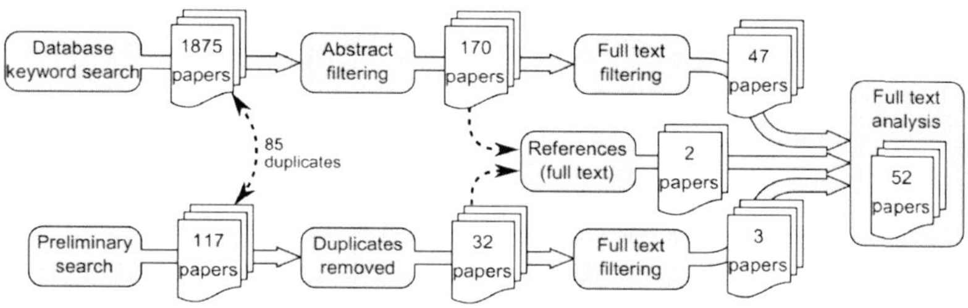

Figure 2: The study selection process.

Table 3: Facets and related search terms used.

Facet	Keywords
Agile methods (before & after)	agile, scrum, "extreme programming", waterfall, "plan-driven", RUP
Organizational transformation	transform*, transiti*, migrat*, journey, adopt*, deploy, introduc*, "roll-out", rollout
Only software related articles	(software OR (conference="agile, xp, icgse, icse")) AND NOT (title+abs="manufacturing" OR conference="agile manufacturing")

Preliminary searches

Before proceeding with identifying the primary studies a number of preliminary searches were performed. The purpose of the preliminary searches was to develop and evaluate different search strings. In addition, we used the searches to identify a set of relevant papers that should be matched by the actual search. We started by examining top ranked hits by trivial keywords that the more complex final search string might miss. Initial searches were made using keywords that were as general as possible, including "agile transformation" and "large scale agile". Based upon these preliminary searches, we selected 117 papers that seemed relevant based upon the title. We used this set as a "sanity check" when developing the actual database search.

Identification of primary studies

The gathering of potential primary studies was based on a search in the online databases listed in Table 2. We constructed a search string based on the previously discussed facets. However, our preliminary searches showed that picking keywords with good precision was difficult. In particular, we did not succeed in representing the facets *large scale* and *empirical* with precise words. Therefore, we included only the facets *agile software development* and *organizational transformation* in the keyword search, with the consequence of increasing the manual filtering effort in the subsequent steps. The used keywords are shown in Table 3.

Study selection

The set of potential primary studies identified by the database search was refined in two steps, first by filtering based on abstracts and finally based on the full text. The study selection process is outlined in Fig. 2.

The database keyword search matched 1875 unique papers. The abstracts of these papers were categorized independently by the two first authors into three categories: include, exclude, and uncertain. There were 1578 exclusions and 62 inclusions that both researchers agreed on. The inclusion decisions for the 235 abstracts with disagreement or uncertainty were resolved through discussion. At this stage papers were excluded only if both researchers deemed it clearly irrelevant, including any uncertain cases for full text filtering. As a result 170 papers were selected for full text filtering. Full text filtering was performed by evaluating the text of each article against the four facets of the inclusion criteria.

Filtering was done in two steps. In the first step, the first author extracted data relevant to the four facets. Based on the extracted data, 76 papers could be immediately deemed as included or excluded. The remaining 94 papers were evaluated against the inclusion criteria by the two first authors, and a decision was made after discussing each paper separately. In difficult cases, the third author was consulted to reach a decision. As a result of the full text filtering, 47 papers were selected for inclusion.

We evaluated the result of the database keyword search against the benchmark created in the preliminary search step. We concluded that 85 of the 117 preliminarily selected papers were matched by the database keyword search. The missed 32 preliminary papers were examined, resulting in including three additional papers as primary sources.

In parallel with the full text filtering step the references of all 170 papers selected for full-text filtering were also examined for relevance. Most papers used references very scarcely, typically referencing well known descriptions of agile methods. This step led us to include two additional papers in our full text analysis.

Table 4: Context information extracted from primary studies.

Contextual code	Explanation
Agile method	Agile methods used in the organization (e.g. Scrum, XP)
Business area	The business area in which the organization operates.
Organization size	Mentioning the size of the case organization.
Time of transformation	When has the transformation been in progress, how long has the transformation taken.
Research process	The paper describes a research process.
Geographical location	Where the organization is located.
Large scale definition	A definition of large-scale software development.
Multisite / GSD	Mentioning of a multisite organization.

Table 5: Code families, codes and quotations.

Code family	Description	Examples	Codes	Quotations
Reason to change	Reasons to start the transformation	Demand for faster delivery	30	123
Transformation process	Statements describing the transformation process	Top-down, big bang, step-wise	16	580
Challenges	Statements presenting challenges in the transformation	Change resistance	40	323
Success factors	Statements presenting success factors in the transformation	Management support	44	260
Investing in change	Factors presenting how the organization is investing in the transformation	Training, consultants, tools	5	137
Practices	Practices used or established during the trans- formation, but described as neutral wrt. to successes and challenges	Coaching, piloting, continuous integration, communities of practice	11	170
Contextual	Contextual codes defined in Table 4	Agile method, organization size, large-scale definition	8	215
		Total	154	1575

As a result, we selected a total of 52 papers for inclusion in the analysis stage.

Handling of duplicate reports on a single case

In several cases, more than one primary study discussed the transformation of the same organization. Duplicate descriptions typically focused on different aspects. For example, one paper would highlight the viewpoint of the developers (Fry and Greene, 2007), and another would consider the transformation from the user experience designers' point of view (Federoff and Courage, 2009).

Even if the transformation of a single organization was described in many studies, all sources that passed the inclusion criteria were included. Studies presenting the same organization were treated as one unit so that we could gain as much insight on each organization as possible. Conversely, there were also a few papers that presented multiple case studies, and in those cases we treated each studied organization individually.

Instead of using the most complete paper as suggested in the guidelines for systematic literature reviews (Kitchenham, 2007), we combined the results presented in each paper and considered the case as a single unit in our analysis. Keeping in mind the potential bias caused by duplicate publications, we think that including all papers enabled us to get a more in-depth understanding of the individual cases.

As a result, we identified 42 unique organizations in the primary studies. We use the term *study* to refer to the primary publications, and the term *case* to refer to an individual case organization that may be described in several different studies.

Study quality assessment

The primary studies for this literature review are almost exclusively industry experience reports. We identified only six case studies with a clearly defined research method, and observations on transformation were presented only as a minor part in these studies. Based on this finding we conclude that case studies presenting insights into large-scale agile organizational transformations are very scarce. We deemed that the results would be distorted heavily and many valuable studies would be left out if a strict quality assessment would be part of the inclusion criteria.

As a result we decided to include all experience reports, regardless of the potential problems of author and publication bias.

Coding of primary studies

We coded the primary studies using an integrated deductive and inductive approach, coding both a contextual and a findings part (Cruzes and Dyba, 2011). We established an a priori list of codes for contextual information (Table 4), which included the codes agile method used, business area, organization size, time of transformation, research process used, geographical location, definition of large-scale used in the paper, and multisite case. Codes related to the research questions were created by an inductive process to avoid having our previous assumptions affect the choice of codes.

To guide inductive coding, we established five different code families a priori: reasons to change, transformation process, challenges, success factors, and contextual. The four first families were related to our overall research questions. In addition to these five evident families two additional families were established during the coding process: investing in change, and practices. A description defining the scope and containing/ examples of codes were defined for each family. The example codes would not necessarily be used as actual codes, and would not represent the entire or final scope of the families. Table 5 presents the codes families in the final state of the coding. Table 5 also shows the total number of codes and quotations created in the coding process. The total number of quotations is less than the sum of quotations in the categories because a single quotation may have multiple codes. Note that in this paper, we only discuss the results related to challenges and success factors, and that it as such forms a part of a larger study.

Synthesis of primary studies based on coding

We synthesized our findings by creating an initial organization of codes into high level categories based on the code labels. Each code was classified into a single category.

After assigning each code to a unique category, the content of the categories was studied. Each category was studied by reading through each quotation of each code included. Typically the quotations were displayed and examined in their original context, considering the surrounding paragraphs of the quotation. Notes were made on each quotation presenting noteworthy observations. We used the notes to create a concise description for each code.

We then refined the categories based upon the concise descriptions. Some codes were re-categorized, and the definitions of a few categories were revisited, until a final ordering was reached. The results are presented according to the final categorization.

The tables presenting the categories of challenges and success factors list references to the primary studies and case organizations. We have included a count of how many individual cases mentioned each factor, and a percentage of this count relative to the 42 individual cases. The purpose of these counts is to give the reader a sense of how often the factor was mentioned. We want to note that merely the frequency of mentioning does not make as generalizable evidence for the actual importance of the factors.

RESULTS

In this section we present our findings. First, we present an overview of the primary studies. Then, we discuss the findings organized according to our research questions.

Overview of the primary studies

In this section we present the type of the primary studies, some characteristics of the case organizations, as well as the agile methods used by the cases.

Study types

The results include findings from 52 primary studies discussing how 42 different software organizations introduced large-scale agile methods. Most studies were experience reports (46 studies). Only six of the included studies had a research method explicitly stated. The publication forums of the primary studies were distributed so that 47 sources were conference proceedings, four sources were journal articles, and one source was a technical report.

In most cases it was evident that the author had been personally involved in the transformation. In 41 out of the 52 papers one or more authors had a direct affiliation to the case organization. Table 6 summarizes the affiliations of the authors. Internal affiliation refers to cases where the author is employed by the case organization. In some cases there are multiple authors with different affiliations.

Table 6: Author affiliations.

	Affiliation type	**Papers**
Experience reports	Internal to organization	32
	Consultant and internal to organization	7
	Consultant	5
	Researcher	2
Case studies	Researcher	4
	Researcher and internal to organization	1
	Internal to organization	1

All included studies and transformations were dated after year 2000. There were peaks in transformation studies in 2008 and 2011. The studies were typically published two years after the start of the transformation.

Tables 7 and 8 list the case organizations and primary sources. Some key contextual information on the organizations is included in the tables. We reference the primary studies with a capital P and a number, and the unique case organizations with a capital C and a number.

Table 7: Experience reports

Case	Paper(s)	Company	Business area	SW org size	Initial state	Agile methods	Trans- from start	Relevance
C02	P03	Amazon.com			Small autonomic teams	Scrum	2004	5
C03	P04	BEKK consulting	IT- and management consulting	160	Waterfall	Unified Process	2003	3
C04	P05, P17	BMC Software	Distributed Systems Management business unit	300	Waterfall, traditional, dispersed	Scrum	2004	5, 5
C05	P06, P09	Yahoo!	Online services	150 teams	Waterfall (named PDP), gated, city-states	Scrum, Lean	2005	4, 5
C06	P07, P52	Nokia Siemens Networks		500		RAD, Scrum	2007	3, 2
C07	P08	ABC Bank	Banking	300		Scrum, UP, XP. Custom delivery		3
C08	P10	Yahoo! Music	Media player for consumer		Enforced top-down, heavy-weight, BDUF, waterfall	Scrum	2006	4

C09	P11	Unisys Cloud Engineering			Waterfall	Scrum	2010	5
C10	P12	BBC, iPlayer	Media	10 teams	Independent processes, Heavy practices	Scrum		4
C11	P13, P15, P16, P28, P33	Salesforce.com	Online services, SaaS CRM software	30 teams	Loose waterfall-based process	Scrum, XP, Lean, ADM	2006	2, 2, 5, 2, 2
C12	P14	MyBoeingFleet.com	Extranet website		CMMI, Macroscope, gate-based	Scrum	2007	3
C13	P18	Ericsson Netherlands	Telecommunications	300		Scrum, XP, Lean	2007	4
C14	P19	Citrix Online	Conferencing and screen sharing software	44 teams	RUP waterfall	Scrum	2007	1
C15	P20, P21	Anonymous	Government	50	Waterfall	Scrum	2010	3, 3
C16	P22	Ericsson R&D Finland	Telecommunications	400	Traditional, functional, silo-based	Scrum, Kanban, Lean		2
C17	P23, P32	British Telecom	Telecommunications	14000	Old fashioned process- driven; Waterfall	Mentioning of Scrum, XP	2004	3, 2
C18	P24	SoftwarePeople			CMMI level 3	Scrum	2006	4

C20	P26	Nokia	Telecommunications		Waterfall, some Scrum on team level	Scrum	2007	2
C21	P27	Anonymous	Healthcare	300	Waterfall	Scrum, XP	2005	3
C22	P29	Microsoft IT	IT services	9 teams	Waterfall and PMI methodologies	Scrum	2006	3
C23	P30	Healthwise	Healthcare, DSS system	200	Expanding organization	Scrum, Lean	2004	3
C24	P31	Borland		300		Scrum	2006	4
C25	P34	Siemens Medical Solutions USA	Health services	300		Scrum		4
C27	P36	Anonymous	Financial services	100	Traditional, phase-based approach	XP	2001	4
C28	P37, P38	Gale	Online services	6 teams	Waterfall, overall unclear		2009	3, 3
C29	P39 (T1)	Anonymous	Online services	50	Waterfall	Scrum		3
C30	P39 (T2)	Anonymous	Banking	9 teams		Scrum		3
C31	P40	Quintiles Trans-national Corp			Homegrown waterfall, CMMI, silos	Scrum		3
C32	P41	Nyx		270	Traditional waterfall-style (RUP)	Scrum	2007	4

C33	P43	VeriSign Enterprise Security	Information security	50		Scrum	2006	4
C34	P44	Primavera Systems	Enterprise project portfolio management	90	Waterfall	Scrum	2003	3
C35	P45	SAP AG	Enterprise applications	18000	Waterfall-like process	Scrum, Lean	2006	4
C36	P46	KeyCorp	Financial	1500	Waterfall, command-and-control	Scrum	2005	5
C37	P47	Capital One				Scrum	2004	3
C38	P48	Cisco Voice Technology Group	Telecommunications	1500	Waterfall	Scrum		4
C39	P49	Qwest Communications	Telecommunications	5000	Rigorous, collaborative in one unit, CMM	XP	2002	5
C40	P50	Pegasystems	Business process management	200	Traditional project management	Scrum	2009	3

Table 8: Case studies.

Case	Paper(s)	Company	Business area	SW org size	Initial state	Agile methods	Transfrom start	Relevance
C01	P01, P02	Anonymous, HQ in the UK	Telecommunications	7500		XP	2005	3
C16	P42	Ericsson R&D Finland	Telecommunications	400	Traditional, silo-based	Lean, Scrum	2009	2
C19	P25	Nokia Siemens Networks	Telecommunications	150	Plan-driven, waterfall	Prac-tices (XP), Scrum	2010	1
C26	P35	Anonymous		50	Waterfall	XP		2
C41	P51 (C1)	Anonymous	Messaging services	100		Scrum	2008	2
C42	P51 (C2)	Anonymous	Facility management	300	Prince2, Wa-terfall	Scrum	2004	2

Table 9 summarizes the content of included research papers. Whereas none of the studies had their research focus directly on the transformation process, they provided similar information as shallow experience reports to our study.

Table 9: Summaries of case study findings.

Case	Study focus	Key results the study focuses on	Subjective relevance	Source material
C01	The effects of agile transformation.	Elements of interpretation and practice.	Focus not on the transformation process.	7 interviews, observed 4 meetings and two days an agile team, documents
C16	How Lean thinking is implemented in software development companies	The current state of practice reflected against a Lean framework	The focus mostly on the current state instead of the transformation process.	17 focus group sessions, material walkthrough workshop, process documents. Sessions involved 21 company representatives.
C19	Has visibility, reaction speed, quality, and motivation changed; comparing before and after	Changes in indicators during the transformation	Focus not on the transformation process.	Defect data, survey responses, comments from feedback sessions after the surveys
C26	Good and bad aspects of communication, during an agile transformation	Brief narratives from a multitude of perspectives evaluating the transformation	Relevant but very brief narratives providing some insights on the transformation process.	Survey questionnaire including open ended questions. Further clarifications via Skype or email.
C41, C42	Multiple-case study on analyzing Scrum introduction paths.	Brief narrative through the steps preparation, introduction, and customization.	Quite relevant, but very brief one paragraph narrative.	3-5 interviews per case.

In the above mentioned tables we have evaluated the relevance of each primary study with respect to our research questions with a subjective score. We defined the scores as follows. 1: some sentences revealing factors relating to transformation. 2: a couple of paragraphs describing transformation. 3: Several paragraphs throughout the paper explaining transformation. 4: entire sections giving a narrative of the transformation. 5: the entire paper focuses on describing how the transformation proceeded.

Case organizations

The size of the organizations varied from the minimum included size of 50 to 18,000 people. The median size was 300 people. In seven studies the size was presented in terms of teams, ranging from the minimum of six teams to 150 teams. The median was 10 teams. In eight studies there was no direct indication about the size but the issues discussed revealed that the organization was of a large size according to our definition.

Different business areas were represented quite evenly. Online services was the largest group, including companies providing software as a service solutions for businesses, online media players for consumers, online services for consumers, and communication software for businesses. The second largest group was telecommunications, including companies such as British Telecom, Cisco, Ericsson, and Nokia Siemens Networks. The third largest group was enterprise management solution providers with products for business process management, project portfolio management, and facility management. Table 10 summarizes the business areas and maps them to cases.

Table 10: Business areas.

Business area	**# of cases**	**Case reference**	**Primary source reference**
Online services for consumers and business	9	C5, C8, C10, C11, C12, C14, C28, C29, C41	P6, P9, P10, P12, P13, P15, P16, P28, P33, P14, P19, P37, P38, P39, P51
Telecommunications	8	C1, C13, C16, C17, C19, C20, C38, C39	P1, P2, P18, P22, P23, P26, P32, P25, P48, P49

Enterprise management	5	C3, C34, C35, C40, C42	P4, P44, P45, P50, P51
Banking	4	C7, C27, C30, C36	P8, P36, P39, P46
Healthcare	3	C21, C23, C25	P27, P30, P34
IT services	2	C4, C22	P5, P17, P29
Government	1	C15	P20, P21
Information security	1	C33	P43
Not specified	9	C2, C6, C9, C18, C24, C26, C31, C32, C37	P3, P7, P52, P11, P24, P31, P35, P40, P41, P47

Agile methods used

The most prevalent agile method used in the transformed organizations was Scrum, which was the sole agile method mentioned in 25 cases. The second most mentioned agile method was Extreme Programming (in 4 cases), which was often combined with Scrum (in 5 cases). Lean software development was mentioned in 6 studies, although in all cases in combination with Scrum. Other agile methods mentioned were Unified Process[1], Adaptive Development Methodology (ADM), and Rapid Application Development. In one case the agile method was not named.

It was quite common that organizations sought to combine agile methods. Especially Scrum, XP, and Lean software development were used together. In addition, many cases mentioned use of XP practices, such as test driven development and continuous integration, without explicitly stating XP as the process being used. Combining and customizing agile practices was also evident as many organizations viewed the agile method as continuously evolving. Organizations evolved the agile methods for instance through retrospectives and continuous improvement.

Transformation challenges

Any organizational transformation that involves numerous individuals will face challenges. In this section we answer our first research question, RQ1: *What challenges have been reported for large-scale agile transformations?*

We organized the found 35 challenges, each mentioned by several sources, into nine categories. The categories are summarized in Table 11.

Table 11: Challenges.

Challenge type	Primary sources	Case organizations	# of cases
Change resistance 16 (38%)			
General resistance to change	P14, P18, P22, P44, P45	C12, C13, C16, C34, C35	5 (12%)
Skepticism towards the new way of working	P5, P8, P9, P18, P29, P36, P46	C4, C5, C7, C13, C22, C27, S36	7 (17%)
Top down mandate creates resistance	P2, P12, P37, P49	C1, C10, C28, C39	4 (10%)
Management unwilling to change	P2, P3, P6, P49	C1, C2, C5, C39	4 (10%)
Lack of investment 13 (31%)			
Lack of coaching	P1, P6, P23, P45, P47, P49	C1, C5, C17, C35, C37, C39	6 (14%)
Lack of training	P11, P20, P45	C9, C15, C35	3 (7%)
Too high workload	P37, P42, P49	C16, C28, C39	3 (7%)
Old commitments kept	P11, P21	C9, C15	2 (5%)
Challenges in rearranging physical spaces	P6, P14, P29, P36, P38, P49	C5, C11, C22, C27, C28, C39	6 (14%)
Agile difficult to implement 20 (48%)			
Misunderstanding agile concepts	P6, P26, P37, P38, P42, P44, P45, P48, P49	C5, C16, C20, C28, C34, C35, C38, C39	8 (19%)
Lack of guidance from literature	P5, P6, P10, P13, P21, P27, P30, P45, P49	C4, C5, C8, C11, C15, C21, C23, C35, C39	9 (21%)
Agile customized poorly	P9, P29, P44, P46	C5, C22, C34, C36	4 (10%)
Reverting to the old way of working	P6, P11, P12, P18, P21, P38, P44, P49	C5, C9, C10, C13, C15, C28, C34, C39	8 (19%)
Excessive enthusiasm	P4, P6, P12, P20	C3, C5, C10, C15	4 (10%)
Coordination challenges in multi-team environment 13 (31%)			

Interfacing between teams difficult	P9, P10, P13, P17, P24, P26	C4, C5, C8, C11, C18, C20	6 (14%)
Autonomous team model challenging	P7, P13, P29, P51	C6, C11, C22, C41	4 (10%)
Global distribution challenges	P29, P45, P48	C22, C35, C38	3 (7%)
Achieving technical consistency	P5, P24, P26, P27, P50	C4, C18, C20, C21, C40	5 (12%)
Different approaches emerge in a multi-team environment 9 (21%)			
Interpretation of agile differs between teams	P8, P9, P38, P43, P48	C5, C7, C28, C33, C38	5 (12%)
Using old and new approaches side by side	P1, P8, P10, P26, P29	C1, C7, C8, C20, C22	5 (12%)
Hierarchical management and organizational boundaries 14 (33%)			
Middle managers' role in agile unclear	P2, P11, P27, P30, P38, P49, P52	C1, C6, C9, C21, C23, C28, C39	7 (17%)
Management in waterfall mode	P3, P6, P35, P38, P45, P46	C2, C6, C26, C28, C35, C36	6 (14%)
Keeping the old bureaucracy	P20, P48	C15, C38	2 (5%)
Internal silos kept	P10, P11	C8, C9	2 (5%)
Requirements engineering challenges 16 (38%)			
High-level requirements management largely missing in agile	P24, P31, P39, P45, P51	C18, C24, C29, C35, C41	5 (12%)
Requirement refinement challenging	P1, P17, P33, P48	C1, C4, C11, C38	4 (10%)
Creating and estimating user stories hard	P5, P11, P27, P30, P33, P37	C4, C9, C11, C21, C23, C28	6 (14%)

Gap between long and short term planning	P9, P10, P13, P26, P31, P38	C5, C8, C11, C20, C24, C28	6 (14%)
Quality assurance challenges 6 (14%)			
Accommodating non-functional testing	P17, P28, P31, P48	C4, C11, C24, C38	4 (10%)
Lack of automated testing	P10, P11, P17	C4, C8, C9	3 (7%)
Requirements ambiguity affects QA	P5, P48	C4, C38	2 (5%)
Integrating non-development functions 18 (43%)			
Other functions unwilling to change	P1, P2, P5, P9, P10, P17, P18, P28, P31, P38, P42, P45, P46, P48, P50	C1, C4, C5, C8, C11, C13, C16, C24, C28, C35, C36, C48, C40	13 (31%)
Challenges in adjusting to incremental delivery pace	P6, P9, P10, P15, P17, P28, P46, P48, P50	C4, C5, C8, C11, C36, C38, C40	7 (17%)
Challenges in adjusting product launch activities	P31, P38	C24, C28	2 (5%)
Rewarding model not teamwork centric	P3, P6, P38, P40, P42, P49	C2, C6, C16, C28, C31, C39	6 (14%)

Change resistance

General resistance to change. People are not willing to change unless there are good reasons that they understand, and the change is perceived easy enough. It is difficult to attain a buy-in for a change, and even organizations that have a flexible culture will face resistance [P10]. It should be expected that not everyone will be willing to change, even so that some employees will never adapt to the new way of working [P6]. During a transformation period, objections to change may become a major reason for loss in time and productivity [P29].

Numerous reasons for change resistance were reported. For instance, Fecarotta [P14] described the organization of Boeing as risk-averse and cautious. Any change was further hindered by a deeply rooted status quo and high employee retention [P14]. In some cases it was reported that people worried about new roles and responsibilities that agile might bring [P44]. For instance, testers were worried that they would need to take on cross-functional tasks, which would be outside their area of expertise [P18]. Yet another reason for change resistance was the move from individual offices to team areas [P22]. People felt that they were being monitored more because of the increased level of interaction within the team and between the various project stakeholders [P45].

Skepticism towards the new way of working. Skepticism and distrust in agile development in general were other common problems. Seffernick describes how management did on the one hand acknowledge the benefits of agility, but on the other hand opposed its introduction due to contract reasons, the current matrix organization, and other organizational practices [P46]. Lewis reports that tending to skeptical team members wasted time and effort [P29]. Skepticism was often created by misconceptions, including that agile does not work for complex products [P5, P36], agile needs to be implemented in a prescriptive by-the-book way [P9], that frequent meetings will cause overhead [P18], and that agile equals avoiding governance and working without a plan [P8].

With skepticism we refer to any kind of reservedness towards the new way of working. While reservedness might not be a problem on its own, it is still worth to acknowledge that reservedness is mentioned in many cases.

Top down mandate creates resistance. The way the transformation is initiated affects how change resistance will show. In several cases, the change initiative came from management, and when presented in a bad way, became a mandate that people were not receptive for. Spayd [P49] summarizes this as: *"Organizations do not change merely because the boss says so, at least not in the way that is intended"*. In that organization the introduction of CMM had been carried out consistently by a mandate, but the collaborative nature of agile did not mix well with such a mindset. A top-down mandate may also dilute the understanding of the reasons for starting the transformation and the understanding of agile development overall [P2]. O'Connor describes how team members became skeptical towards agile when implementing the mandated changes did not bring any visible benefits [P37]. A problem relating to a top-down mandate is that if management does not define a clear goal for using agile, developers may feel that the agile methods may be replaced by something else at any time [P12].

Management unwilling to change. Change resistance amongst managers can also create problems. In cases where the change was emerging bottom-up, management became reluctant to change, making significant organizational change above the team level impossible. For instance, Atlas [P3] describes how executive support would have been required to extend the agile process to involve the product management office and separate quality assurance groups. In this case, when managers were not involved in the transformation, the agile way of working could not spread beyond the development teams [P3]. Lack of middle management support for change and a disinclination to change management culture were seen as some of the most serious problems in the transformation [P2, P49]. Middle management is in a position to undermine the entire transformation, and may do so if they do not participate in, and thus understand, the agile method. Agile brings changes to some management roles [P2], and lack of understanding of agile development will leave managers feeling left out [P6].

Lack of investment

Lack of training. Training and coaching are direct investments in transformation and their lack is an evident problem. Not providing enough funding for these activities can create difficulties in the transformation [P11]. Hajjdiab [P20] describes how reluctance of management to invest in training caused teams to be ill prepared for the transformation effort, eventually resulting in ending the use of agile methods. A less dramatic outcome of too little training was lower motivation [P45].

Lack of coaching. Arranging proper coaching was a problem in many organizations. It is critical to coach teams in their real work environment as a proper change in mindset is difficult to achieve only by attending training sessions [P9, P23].

In cases where numerous teams needed to be coached the demand often exceeded the capacity of available coaches [P1, P6, P23, P49]. Lack of coaching was also attributed as a reason why the success of pilot teams could not be repeated when agile was adopted more widely [P9, P45]. Silva and Doss [P47] described that when it was hard to fill the coaching positions, people less seasoned in agile were appointed as coaches, which created the risk that agile practices would not be taught correctly.

Too high workload. Even though the case organizations were in a state of transformation, the workload of the teams was not always adjusted to facilitate the change process. Some organizations started their agile journey in a state where people were over-committed, and later realized that overloaded people will not be able to change their behavior and learn new ways of working [P37, P42].

Old commitments kept. Sometimes all old commitments were kept despite the transformation. In one case, management forced people to remain committed to firm deadlines, even midst in the organizational transformation [P21]. The engagement in old commitments and tasks resulted in ignoring new agile practices, and eventually the agile method broke down [P21]. One case [P11] describes how time was allocated for tending to old commitments. However, even though they had prepared for extra work, people with specific expertise became overloaded.

Another case describes how senior management pressed teams to deliver what the customer had been promised, regardless of what the current velocity of development predicted. The pressure to deliver interfered with the transformation, but also forced a change in the old way of working [P49].

Rearranging physical spaces. In some cases, the old way of working had people seated physically distributed, in opposition to what agile methods suggest. Office spaces needed to be arranged so that the entire team could work in a single room, but it took time and effort, and it was not always possible [P6, P14, P36, P38]. Lewis and Neher describe how dedicated rooms for teams could not be arranged, and team events such as daily stand-ups required booking a conference room [P29]. In another case, the facilities organization shut down the teams' attempts to modify their working spaces [P49].

Agile difficult to implement

A common challenge was that implementing agile methods turned out to be difficult. An experienced software team may do well in training, but when the time comes to apply agile techniques in practice, the team may get lost [P21].

Misunderstanding agile concepts. There were many examples of problems caused by misconceptions of what agile software development is. On a general level, Bang [P4] describes how the values of the agile manifesto were not understood, and agile practices were carried out without understanding their purpose. In one case, management saw the purpose of agile simply being faster product delivery [P9].

Examples on the team level include how agile was seen as the freedom to hack without documentation [P26], that development could be done without design tasks [P37], and that agile meant that everyone should become a generalist [P42]. Schatz and Abdelshafi describe how teams presented unfinished work at reviews and ignored the principle of showing only completed increments, which resulted in a backlog of bugs [P44]. In one case, team members had perceived the introduction of agile as a means to squeeze out more efficiency [P45]. In addition, the flatter organization was seen as fewer career opportunities, pushing team members to compete for visibility [P45].

A further misunderstanding of agile was due to viewing it only through the tools used, such as management software [P48]. Superficially focusing only on the tools themselves and not the reasons behind their use led to frustration [P42].

In some cases the misunderstanding of agile led to *"doing mini waterfalls"* [P6]. Managers used Scrum terminology, but had the teams commit to unreasonable workloads [P6]. In another case, a waterfall nature was evident as task estimates were given as hours left, and task breakdowns became disconnected from what was really being done [P27].

As a final example of misconceptions, some organizations considered agile being a solution to all problems [P9]. Success was declared prematurely [P49] but expectations created by successful early experiments were not fulfilled [P38].

Lack of guidance from literature. Several cases describe how an agile method was hard to learn from the literature [P21, P27], especially when it comes to implementing it on a large scale [P13]. As Beavers [P5] writes: *"There simply was not a manual or document where we could find easy answers on how to do things"*. Schnitter and Mackert [P45] report that theoretical considerations on how to scale up the agile methods were not good enough, and that product managers and architects struggled when several Scrum teams were working concurrently. Another case concluded that all practices suggested by XP did not fit enterprise scale development, and some practices required customization [P49].

Benefield [P6] describes how it was difficult to find a balance between prescribing a by-the-book implementation which may put people off, and giving too much freedom in the agile methods, which may weaken core practices.

The reality where the practices needed to be applied was described as messy in comparison to the ideal circumstances presented in the literature [P10]. Thus, it was difficult to choose a method to start with and gain buy-in for [P10]. Difficulties in choosing an initial approach were also due to variances perceived in the agile approaches suggested in literature [P30].

Agile customized poorly. Furthermore, the difficulty of and misunderstandings related to agile were evident in cases where the methods were customized poorly. As a by-the-book implementation often was not feasible, organizations attempted to tailor the agile method to suit their specific needs. However, in some cases this simply meant skipping practices, which led to problems. In one case, certain individuals were allowed to ignore core elements of Scrum, which turned the teams' decision making into a variant of command and control [P9].

In one case [P29], there was a temptation to strip some agile practices and enhance others. Previous attempts had proven that one of the reasons for agile implementations to fail was deviations from the process, because of which the agile mindset did not take root [P29]. A poor customization may lead teams to adopt only practices that reflect their current needs, thus failing to achieve any real change in process and mindset [P9, P46].

A further case of poor customization included replacing the agile vocabulary with familiar terminology from the old approach [P44]. An important benefit of introducing new vocabulary is that it stimulates new ways of thinking [P44].

Reverting to old way of working. In several cases, challenges in the transformation resulted in people reverting to the old way of working. In some cases it was only a temporary struggle while learning agile practices, but in other cases the old way of working displaced agile.

Development work has to go on during the transformation but there will be new things to learn for the team. Stress caused by the combination of schedule pressure and much change at once can pull people back to the old way of working [P11, P18, P38].

Even subtle trouble may put the transformation at risk, as people will always look for reasons to revert to familiar behavior [P44]. Teams without adequate training were struggling with applying agile practices correctly, and the challenge the new practices posed made people abandon them and return to the ways they know [P6, P49].

In some cases agile was seen as an overhead, and was therefore abandoned. For instance, as new practices were being introduced there was a decrease in performance, and when teams realized that the benefits are not immediate, they started reverting to the old way of working [P21].

Evans [P12] describes that new senior appointments made management less favorable towards agile, and a waterfall development method started to return.

Excessive enthusiasm. A phenomena that troubled some organizations was over-enthusiasm towards agile methods. Several cases contained reports of change leaders or teams becoming agile zealots. For instance, some members of the agile community could become too evangelic, making people take sides for or against agile development [P12]. Also, while starting out with an overly eager attitude, team members' interest faded, and they reverted to old ways of working when the new approach did not deliver immediate benefits [P20].

Attempting to implement agile too strictly may cause conflicts, and especially when implementing large-scale agile, the change leaders need to maintain a collaborative attitude towards various groups in the surrounding organization [P6]. Introducing agile methods does not guarantee success, and therefore they should not be followed blindly [P4].

Coordination challenges in multi-team environments

Interfacing between teams difficult. One of the most prominent transformation challenges was the difficulty in coordinating the work of several agile teams.

Challenges arose when teams needed to work with other teams, and as parts of the larger surrounding organization [P13, P17, P24]. While introducing agile had created flexibility at the team level, the surrounding organization was not responsive enough [P26]. The roll-out of agile had not removed dependencies, and the dependencies made managing development difficult [P9, P10].

Autonomous team model challenging. Some organizations initially created models in which teams operate autonomously — well in line with agile principles. However, a number of issues arose from this independence. For instance, teams needed to strike a balance between the their own goals and the broader goals of the organization, but often chose to focus only on their own goals [P7]. Coordination was obstructed by independent teams that did not respect the larger context [P13]. Coordination was also difficult when teams were working independently for different customers, but the applications being built were interdependent [P29]. One case reported that even allowing teams to have different Sprint durations created delays in delivery [P51].

Global distribution challenges. Further coordination problems were encountered when scaling up agile over many geographically distributed sites. Distribution had negative effects, such as missing kick-off meetings, reduced feelings of proximity when telecommunication is necessary, and difficulty in arranging frequent meetings due to time zone differences [P45]. Agile project management was also problematic. In a waterfall way of working, separate parts of projects could be isolated to different sites, but the agile way of working does not allow such strict slicing of projects [P29]. In one case, it was simply admitted that a distributed organization will impose additional burden on communication and require additional care in the process, but distribution and agile would still be used together [P48].

Achieving technical consistency. Technical problems relating to inter-team coordination were reported as well. Integrating the products of teams was problematic in some cases [P5, P50], and the lack of standardized build scripts in one case [P27]. There were also problems in synchronizing the definitions of software interfaces between teams, and dependencies in code caused problems in larger features [P26]. One case reported that the strong focus on individual teams in the agile way of working created a fragile architecture, divergence in coding style, and even distrust between teams [P24].

One case describes the progression of coordination problems. First, an attempt was made to reduce cross-team dependencies, but it became evident that the dependencies were an inherent part of the project. A traditional approach to managing dependencies created excess work, reduced independence and flexibility of teams, and created contract-based and adversarial relationships. Coordinating using Scrum-of-Scrums worked for up to five teams, but did not scale to a global level, as teams were focused on their own work and the new practices narrowed communication channels and created communication breakdowns. It was concluded that a balance is needed between completing new stories from the team backlog and maintaining overall stability of the application. [P34]

Different approaches emerge in a multi-team environment

Interpretation of agile differs between teams. When many teams implement agile without consistent guidance, friction and fragmentation may emerge [P9]. The organization may require moving people between teams from time to time, and therefore it is desirable that the agile cultures of different teams are not too different. Divergence in process creates increased costs when relocating people [P38]. Further, forecasting and benchmarking teams become difficult [P38, P48]. To overcome problems with divergence in agile approaches some organizations defined standards [P8, P43].

Using old and new methods side by side. In most cases, the organizational transformation proceeded gradually, and during the process it was possible that the new agile method was used in parallel with the old methods. Using the old and new methods side by side was generally seen as problematic [P1, P26], causing tensions on all organizational levels [P10]. Ongoing projects had been set up with old development methods, and the agile methods needed to be arranged to fit in with them [P8, P29]. A particular problem in collaboration between waterfall and agile projects was that in agile, the software design was fleshed out over time as sprints progressed, but the waterfall method required a detailed design to be frozen upfront [P29].

Hierarchical management and organizational boundaries

Middle managers' new role in agile unclear. The need for additional management positions in a larger organization may pose problems to agile processes that emphasize self-organization. Especially the role of middle management was unclear in agile methods [P2]. This is problematic, as an agile transformation requires a cultural change particularly on the middle management level [P2]. Managers were reported to need to resist the tendency to command and control and allow room for self-organization, but the change in mindset was difficult to achieve for the people involved [P11, P52]. One case [P30] describes how the project management group had previously worked through big up front plans and competed for resources, but those ways would need to dramatically change.

Several other problems related to management roles were also presented. For instance, [P49] describes how managers were left outside the roles offered by the new agile way of working. In another case, when managers were appointed as Scrum masters, developers felt being micromanaged [P27]. This was partially because of a poor understanding of the agile method [P27]. It was also described how mixing the role of project manager and Scrum master created a conflict of interest, and turned the Scrum manager's role into one of policing instead of supporting the team [P30]. In some cases, problems with manager positions were solved by phasing out roles relating to the old way of working, and replacing them with new positions more suitable for agile development [P38, P52].

Management in waterfall mode. Even after adopting agile, there were cases where management continued to work according to the old waterfall model. One case [P38] described management as *"focused on meetings and big up-front project plans"*, despite having adopted agile. In another case, management was losing confidence in agile because reports on costs and progress were not produced in the same format as before [P35]. As Scrum teams did not commit to fixed schedules, they were considered unreliable [P45].

Some cases reported that development efforts were still controlled on the top level by a project management office (PMO). The PMO was described as entrenched in a waterfall paradigm, and the top-level rigidity was causing friction in the agile adoption [P3, P6]. The PMO was seen as a hub and bottleneck controlling all aspects of projects, and its central structure had to be dismantled for the organization to become agile [P46].

Keeping the old bureaucracy. There were also problems with duplicating bureaucracy when two different ways of working were in effect. For instance, agile teams were required to comply with current procedures producing excess documentation and stepping through approval gates [P20, P21]. The bureaucracy of the previous traditional model was still enforced, and management was not willing to lighten the process. Another case describes that after introducing the agile method teams were required to fill in templates of two processes [P48].

Internal silos kept. In some cases, the initial organization had internal boundaries, or specialized knowledge in silos, causing problems in the agile implementation. Cloke [P10] describes how the use of Scrum revealed an internal segmentation where teams operated with differing priorities and agendas. This hampered the initial effort to use Scrum in a larger context [P10]. Cowan [P11] describes how projects needed specialized skills that were scarce and people were often relocated to match the needs of the moment. Too much reorganization made it difficult for teams to plan ahead [P11].

Requirements engineering challenges

High-level requirements management largely missing in agile. While some agile methods, e.g. DSDM, have a quite structured approach to requirements, e.g. Scrum which was the most widely adopted one does not, nor does, e.g. XP, another widely used approach. Large development projects demand high-level requirements management. This is apparent in cases where product complexity requires additional layers of product management [P45], requirements are created by several stakeholder groups and development teams cannot be in contact with all of them [P39], or do not have access to stakeholders due to distribution [P24].

Requirement refinement challenging. In some cases, requirements were initially defined at a high level in marketing requirements documents [P1, P31] or functional specifications [P33]. These high level requirements needed to be elaborated to be useful by agile development teams. The requirements refinement in itself [P1, P33], when to do it, and to which level of detail [P17] were all reported as challenges.

Creating and estimating user stories hard. Product managers and business analysts struggled with creating high level requirements [P33, P37, P51], and teams struggled breaking them down to a size that is possible to do effort estimation on [P33]. One study [P48] describes how high level product management delivered requirements in large chunks, resulting in development teams spending huge amounts of time defining suitably sized stories.

Several cases highlighted that much learning was needed to master the new process of creating user stories, both on product management and development levels [P27, P30, P33, P37]. There were problems such as ambiguity in requirements [P5, P11], and effort estimation for stories [P27, P37].

Gap between long and short term planning. High level requirements work and estimation was problematic as there was a gap between short and long term planning [P9, P10]. Typical agile backlogs give only short-term visibility, creating a need for additional practices for long-term planning [P26, P31].

Several cases described that care had to be taken to avoid long term planning becoming a prescriptive practice. In order to preserve the agility of development, a schedule-driven approach was avoided [P10], sensitivity was used when considering setting milestones [P13], and striving for more accurate estimates was not allowed to become an excuse for gathering requirements up front [P38].

Quality assurance challenges

Similarly to requirements engineering, the agile approach may need to be extended to accommodate additional testing activities [P31]. Organizations may have existing separate QA teams that must be coordinated to work with development teams [P17, P30, P48].

Accommodating non-functional testing. Full quality assurance of a system might require special testing such as performance, load, and memory testing, but agile methods lack a focus on them [P17]. These testing activities can not be done within the boundaries of user stories, and require more resources than teams can spare [P31]. When testing tasks overlap team boundaries, it is sensible to have separate teams for the tasks, but coordination between specialized QA teams and development teams must be defined [P28, P48].

Lack of automated testing. Several cases reported problems with lack of automated testing. The lack of automated tests caused excess testing work and late discovery of defects [P10, P11, P17].

Requirements ambiguity affects QA. Another challenge in QA was the relationship to requirements engineering. The QA functions were struggling if there were problems such as ambiguity or ineffective breakdown of large requirements [P48]. In the waterfall way of working, an extended time was reserved for QA at the end of projects, which allowed retroactively resolving ambiguities in requirements [P5]. However, ambiguity became an impediment to QA when development worked in short increments and there was no extended period for stabilization [P5].

Integrating non-development functions in the transformation

Several cases described how introducing agile started from the development teams [P2, P17, P31, P38]. However, development operates in a larger context and needs to interface with other organizational functions, causing challenges.

Other functions unwilling to change. Challenges were faced when exposing other parts of the organization to the agile way of working, as functions distanced from development were resistant to change [P2, P18, P31]. Tension grew between development and other roles in the organization [P1]. The full benefits of the transformation could not be attained unless the entire organization was set to work along the same paradigm [P18, P42].

Our primary sources listed various functions beyond development that interface with development and are affected by the agile transformation. Marketing was particularly frequently mentioned [P5, P17, P28, P31, P38, P45, P48]. Other groups included sales [P17, P31], infrastructure and operations [P10, P45, P46, P50], user experience and design [P9, P10, P28], documentation [P28], legal [P10], security [P10], customer services [P48], and finance [P48].

Challenges in adjusting to incremental delivery pace. The iterative nature and the fact that agile affects timings in the software delivery life cycle, often caused problems in the interface between development and other functions.

Iterativity changed the pace of delivery [P17], and forced design to focus on a shorter scope [P15]. Various groups feeding and supporting development had earlier committed once to a long term plan, but in the new arrangement they needed to adjust to incremental deliveries at a faster pace [P48].

Especially user experience (UX) and interaction designers did not welcome an incremental model [P6], and were struggling with maintaining the big picture in design while working in iterations [P9, P10]. The infrastructure and operations teams fell behind due to the increased speed of development teams, and were forced to update their way of working [P46, P50].

Federoff and Courage [P15] elaborate on the user experience team's problems with the short time horizons of Scrum. Previously, the UX team had supplied development with designs within a long release cycle, but Scrum required completing features within weeks. The problem was that the UX work required a holistic view instead of an incremental one, and the design process did not fit well into the time frame of a sprint. The UX team was overloaded and discontent, but the situation was fixed by refining the schedule for interactions with the development teams. [P15, P28]

Challenges in adjusting product launch activities. A second type of timing problem was that certain release activities inevitably have long lead times, and required the commitment to a set of functionality early on. Agile's emphasis on short time horizons and flexibility in prioritization did not mix well with such activities. For instance, preparing printed marketing material and press releases requires information to accurately present the upcoming features of the product [P38].

One case describes that marketing needed three months for preparing the product launch, but the agile process allowed development to keep changing the content of the product. As a result, marketing struggled, creating material and campaigns without knowing the exact product features. Further, the processes of acquiring export clearances and licensing authority required that the feature set of the product was known well in advance. Development felt that these requests slowed them down [P31].

Rewarding model not teamwork centric. In several cases, human resources (HR) was mentioned as working against the agile adoption. In order to gain the full benefits of going agile, the entire organization should to be aligned, including HR [P38, P42, P49]. Especially the rewarding practices set by HR were seen as problematic, as rewards were tied to personal performance, acting against the team-centric thinking and the agile approach in general [P3, P6, P49]. One case even reported a practice of tracing failures down to individuals [P40].

Success factors in transformation

In this section answer our second research question RQ2: *What success factors have been reported for large-scale agile transformations?* We organized the 29 success factors, each identified in several primary studies, into eleven categories, which are elaborated in this section and summarized in Table 12.

Table 12: Success factors for large-scale agile transformations.

Success factors	**Primary sources**	**Case organizations**	**# of cases**
Management support			**16 (38%)**
Ensure management support	P8, P11, P16, P18, P27, P31, P33, P36, P43, P44, P46, P47, P49	C7, C9, C11, C13, C21, C24, C27, C33, C34, C36, C37, C39	12 (29%)
Make management support visible	P9, P11, P24, P40	C5, C9, C18, C31	4 (12%)
Educate management on agile	P6, P8, P9, P11, P29, P46, P47	C5, C7, C9, C22, C36, C37	6 (14)
Commitment to change			**7 (17%)**
Communicate that change is non-negotiable	P11, P22, P48, P49	C9, C16, C38, C39	4 (10%)
Show strong commitment	P8, P10, P11, P43	C7, C8, C9, C33	4 (10%)
Leadership			**7 (17%)**
Recognize the importance of change leaders	P3, P4, P11, P16, P18, P31, P33, P42	C2, C3, C9, C11, C13, C16, C24	7 (17%)
Engage change leaders without baggage of the past	P11, P16	C9, C11	2 (5%)
Choosing and customizing the agile approach			**20 (48%)**

Customize the agile approach carefully	P8, P10, P11, P29, P31, P40, P41, P43, P45, P49, P51	C7, C8, C9, C22, C24, C31, C32, C33, C35, C39, C41	11 (26%)
Conform to a single approach	P8, P11, P12, P16, P32, P43	C7, C9, C10, C11, C17, C33	6 (14%)
Map to old way of working to ease adaptation	P3, P14, P16, P29, P44, P48	C2, C11, C12, C22, C34, C38	6 (14%)
Keep it simple	P5, P10, P16, P40	C4, C8, C11, C30	4 (10%)
Piloting			**14 (33%)**
Start with a pilot to gain acceptance	P3, P9, P14, P17, P31, P36, P38, P39, P47	C2, C4, C5, C12, C24, C27, C28, C29, C37	9 (21%)
Gather insights from a pilot	P8, P22, P27, P40, P45	C7, C16, C21, C31, C35	5 (12%)
Training and coaching			**15 (36%)**
Provide training on agile methods	P3, P6, P14, P16, P30, P37, P46	C2, C5, C11, C12, C23, C28, C36	7 (17%)
Coach teams as they learn by doing	P3, P6, P8, P9, P16, P17, P18, P29, P30, P33, P42, P44, P47, P48	C2, C4, C5, C7, C11, C13, C16, C22, C23, C34, C37, C38	12 (29%)
Engaging people			**9 (12%)**
Start with agile supporters	P6, P29, P34	C5, C22, C25	3 (7%)
Include persons with previous agile experience	P6, P10, P37	C5, C8, C28	3 (7%)
Engage everyone in the organization	P4, P6, P16, P22, P31	C3, C5, C11, C16, C24	5 (12%)
Communication and transparency			**7 (17%)**
Communicate the change intensively	P16, P22, P33, P40, P41, P43, P48	C11, C16, C31, C33, C38	5 (12%)
Make the change transparent	P6, P11, P16, P42, P43	C5, C9, C11, C16, C33	5 (12%)
Create and communicate positive experiences in the beginning	P6, P32, P29, P40, P42, P46	C5, C16, C17, C22, C31, C36	6 (14%)
Mindset and Alignment			**17 (40%)**

Concentrate on agile values	P1, P16, P42, P46, P48	C1, C11, C16, C36, C38	5 (12%)
Arrange social events	P11, P23, P27, P32, P40, P41	C9, C17, C21, C31, C32	5 (12%)
Cherish agile communities	P3, P12, P41, P42, P47	C2, C10, C16, C32, C37	5 (12%)
Align the organization	P8, P18, P31, P42, P43, P46	C7, C13, C16, C24, C33, C36	6 (14%)
Team autonomy			**10 (24%)**
Allow teams to self-organize	P5, P16, P18, P27, P30, P34, P41, P45	C4, C11, C13, C21, C23, C25, C32, C35	8 (19%)
Allow grass roots level empowerment	P3, P6, P41	C2, C5, C32	3 (7%)
Requirements management			**10 (24%)**
Recognize the importance of the Product Owner role	P12, P14, P16, P24, P30, P33, P39, P46	C10, C11, C12, C18, C23, C30, C36	7 (17%)
Invest in learning to refine the requirements	P5, P11, P17, P33, P48	C4, C9, C11, C38	4 (10%)

Management support

Ensure management support. Numerous cases made it clear that management support is an absolute necessity [P8, P18, P27, P31, P33, P36, P43, P47, P49]. This was reflected in statements such as "Adopting agile, or implementing any significant change, requires an executives sincere support." [P44], "Executive commitment was crucial to implementing massive change." [P16], and "Having upper management engaged, supportive and visible is critical for wholesale organization involvement with Scrum." [P11].

Managers were seen to be in a key role in making the change stick, as they had the authority and power to remove impediments [P18]. Seffernick [P46] describes how a number of people attempted to explain away the applicability of agile methods, but the objections were overruled by the director's commitment to make agile work. Management support was similarly needed when tight release schedules had to be flexed in order to give room for the adoption process [P16, P27]. Favorable management decisions were also critical when additional resources were allocated to training and coaching [P47].

Make management support visible. Visible involvement of management was reported to motivate and encourage employees to adopt the new way of working [P40]. For instance, the CTO organized training sessions personally [P24], and the engineering VP frequently visited sprint demos [P11]. When the corporate level support for the agile initiative was showing teams adopted agile methods even spontaneously [P9].

Educate management on agile. In order to gain support for agile several cases highlighted the need to educate management [P8, P47]. Proper education ensured that managers would not enter a command and control mode, which would harm the agile implementation [P46]. Managers not understanding the principles of agile felt left out with the introduction of self-organizing teams, which sometimes resulted in backlashes [P6]. Providing proper training corrected the situation, and even created strong agile supporters in management [P6, P9]. Training cleared misconceptions [P29], and helped create a consistent implementation of the agile approach across the organization [P11].

Commitment to change

Communicate that change is non-negotiable. Feedback from the personnel should certainly be listened to, but in the end it must be made clear that the old way of working can not be returned to [P22, P49].

For a case where the organization was changed all at once it was reported that the magnitude of change helped to create an impression that the change was non-negotiable [P11]. Firstly, as it was evident that a large change was to happen people felt encouragement and permission to move on from the old way, and secondly, the urgency eliminated wasteful discussions on whether Scrum was a good method or not [P11].

When the organization culture is ingrown, the change vision must constantly be reminded of, and each step towards the new way of working considered as a victory [P48].

Transformation can be facilitated by setting up mechanisms that force change. Spayd [P49] describes that senior management pushed hard on a mandate to have deliveries every 90 days. The mandate made change necessary, and while the strong pressure did not promote agile practices in every case, the drive for change was perceived good in general [P49].

Show strong commitment. A large change will inevitably require strong commitment, but problems in the transformation can put the commitment to test. The agile approach is introduced because of problems in the old way of working, and therefore there will be organizational issues uncovered during the transformation [P10]. People must not be demoralized when facing challenges, and the determination to change must be maintained [P8, P10]. Problems might not be due to the agile approach — in some cases it can expose existing problems in the organization [P10].

Ryan and Scudiere [P43] describe that management commitment was showing as a strong focus on certain agile practices that had been chosen as non-negotiable, and constant assessment that the practices work. The expectations were clearly communicated to the teams, and constant assessment made it clear that change was desired [P43]. A strong commitment from management assures teams that the change is the right thing to do [P11].

Leadership

Recognize the importance of change leaders. Transforming the way of working of a large group requires coordination and leadership [P31, P42]. In addition to the leadership provided by coaches, specific change leaders were mentioned. Cases indicated the importance of having spokespersons for the change [P3, P4, P31]. Cowan [P11] describes how one person was strongly driving the transformation, and made an indisputable contribution for transforming the organization. Maples [P31] describes how the change and agile adoption was guided by one "counselor" [P31]. Other cases described that the change was led by a competent roll-out team, which had representatives from all parts of the organization [P16, P33]. Finally, the responsibility of line and project managers to act as change leaders was highlighted [P18].

Engage change leaders without baggage of the past. A few cases discussed the benefit of having change leaders without baggage from the past. Cowan [P11] describes how the newly hired director was able to circumvent territorial battles that existed from before, and therefore focus fully on getting the agile approach implemented. In another case, external coaches were described to better spot places for correction in the agile approach, and their advice was received better because they were considered impartial when being external to the organization [P16].

Choosing and customizing the agile approach

Customize the agile approach carefully. Customizing the agile approach and practices was often seen as a necessary step in the agile implementation. As each organization will have its own challenges in adopting agile, it should be carefully considered which organization specific areas to focus on when choosing the agile practices to implement [P8, P43]. Lewis and Nehrer [P29] recommended choosing the agile approach according to the current corporate model in order to avoid interference.

Cases reported that customization was part of a successful agile implementation. For instance, teams were let to customize their practices individually, to fit the needs of each team [P41]. Spayd [P49] writes that teams who modified agile practices to fit the large and distributed environment ended up doing better than teams that did not.

To allow teams to become innovative and perform well, the agile approach should be customized in a pragmatic way instead of following a strict textbook interpretation [P10]. In order to draw the most out of agile, teams should innovate and find their own practices that really work for their case [P40, P41].

Especially in a larger organization it is not feasible to use the same process for all projects [P49]. An individual application of the agile process is needed for different types of development, depending on, e.g., the type of software being developed or the project size [P45]. Applying agile at scale will require to deviate from some of the typical recommendations [P31]. However, it is essential to remember the agile principles when doing customizations, and watch out for customization that would contradict the principles [P31]. The customization of the agile approach was also seen as an evolutionary process. Cowan [P11] described that in the big bang transformation it was useful to selectively compromise some parts of the agile implementation, so that the core practices were firmly set in place. The agile transformation is a constantly ongoing process, and it is recommended to regularly challenge teams to refine the agile implementation to reflect the current needs of the organization [P41, P43].

Conform to a single approach. Studies reported that conformity to a single approach should be considered when implementing agile. A consistent common vocabulary was seen to benefit the organization and support the change [P8, P16, P43]. Other benefits of using a single approach were the possibility to compare work between teams, easier relocation of people, and predictable progress for the stakeholders [P43]. Consistency and common understanding was created in meetings with peers in the same job discipline, in public events, and by peer coaching [P11, P12, P32].

Map to old way of working to ease adaptation. Mapping the agile approach to the old way of working was seen as necessary in a few cases. Although a general mapping to the old way of may not be good [P44, P46], some cases needed to preserve high-level management practices [P14, P29, P48]. By allowing management to remain unchanged it was possible to introduce agile step-wise on the team level, which helped in getting management buy-in for the process [P14, P29, P48].

In one case, Scrum was considered as a wrapper for existing practices, and could therefore easily fit in the organization [P44]. The agile methodology was considered complementing the existing culture, which eased management buy-in for the new method [P44].

A few organizations had structures in place that were similar to agile, making the transformation easier. For instance, a previous organizational model based on small and autonomous teams strongly aided the adoption [P3]. Another case described the original on-demand delivery model a natural fit for agile, and the transformation was presented as a return to core values instead of a remodeling [P16].

Keep it simple. One piece of advice given by a few cases was to keep the organization and process simple [P10, P16]. Beavers [P5] describes that attempting to operate in a complex and global setting complicated even simple agile practices. The organizational chart was simplified, which was welcomed by both employees and managers [P5]. Rather than focusing on details in process, communication practices, and tools the focus should be on engaging the people in the teams, and forming the process based on what can be seen to work in practice [P40].

Piloting

Start with a pilot to gain acceptance. Having a pilot project was reported as a significant success factor in several cases [P3, P9, P38]. The pilot project helped create confidence that the agile way of working would be suitable for the organization and the general acceptance of agile was increased [P3, P17]. The pilot also cleared disbeliefs of the suitability of large-scale agile, and created acceptance both in development and management [P9, P31, P36].

Pilots were especially important for gaining management acceptance. Cases reported that management gave approval for agile methods only after seeing successful pilots [P14, P47]. Prokhorenko [P39] describes that seeing the example of successful pilots will make managers from other departments eager to try the new method, and thus help to spread its use.

Gather insights from a pilot. A benefit in piloting was that it provided knowledge on how to fit the agile method to the particular organization. A pilot project enabled the organization to get feedback on how teams and managers are best introduced to the new methods [P8, P22]. Organizations met challenges in the pilot projects. However, the pilot projects served as valuable learning experiments providing insights on how to mitigate problems when the rest of the organization is transforming [P27, P40, P45].

Training and coaching

Provide training on agile methods. Several studies stated that training improved the chances of succeeding in the transformation [P3, P16, P30]. It was even highlighted that the change would have failed without training [P14], as Benefield [P6] claimed: *"we saw again and again how training could make the difference between success and failure for a team"*. Training also helped people to become more positively inclined towards the new way of working [P16, P37], and at best training made people enthusiastic to change [P3, P46].

Coach teams as they learn by doing. Agile methods avoid prescribing exact ways of working, and rather emphasize a mindset of adapting to the situation. It is difficult to explain by theory how such a mindset should be applied, and the agile practices are best learned by doing. Coaching teams while they apply the agile methods in practice was seen as an important factor in change [P3, P16, P17, P30, P47]. A few cases stated even that coaching was essential for success in transformation [P8, P18, P29, P33]. Also, without coaching, teams can do damage to the agile transformation if techniques are applied improperly [P9].

There were a number of statements on the benefits of coaching. A coach can watch for and correct problems when they arise [P16, P42]. Coaches helped draw attention away from a focus on tools towards understanding the principles of agile [P42, P48]. There were also benefits in using both internal and external coaches. External coaches were able to have an objective view of the organization [P16, P44], whereas internal coaches were more accessible and knew the specifics of the organization [P3, P6].

Engaging people

Start with agile supporters. Choosing people with a disposition towards agile methods was seen as a requirement for the change to take the right track. Teams are staffed usually based on technical competencies, but also considering personality aspects was emphasized for agile teams [P29, P34]. The personality of people was seen as a key aspect for achieving change [P6]. There is a need for collaborative and understanding persons who are prepared to discard preconceptions and willing to try new untested approaches [P6, P29].

Include persons with previous agile experience. A few cases mentioned the importance people with previous agile experience had at the start of the transformation [P6]. For instance, it helped the product management office to implement agile over the entire enterprise when the staffing was updated to include people with agile experience [P10]. Staffing teams partially with developers familiar with agile helped the rest of the team get a good hold of agile development [P37].

Engage everyone. Many cases highlighted the importance of engaging people broadly in the organization. One of the lessons learned presented was that it is important to involve all stakeholders to gain acceptance of the transformation [P4, P31]. The transition team made an effort to engage people by both inviting them to give feedback online and by holding an extensive number of feedback meetings [P6, P16, P22]. Being inclusive towards everyone was seen as one of the key success factors in the transformation [P16]. Inclusiveness will encourage people to participate and visibly work in the new agile way [P16].

Communication and transparency

Communicate the change intensively. Intensive communication was emphasized in a number of studies. It is important to reach as many people throughout the organization as possible, as without communication the new way of working will not take root [P16, P43]. It was recommended that working in the new agile way is made highly visible on many communication channels and even over-communicated [P16, P33, P41]. Mencke [P33] summarizes the viewpoint on over-communication: *"Even if you think that your teams understand a new method or process, repeat your communication to be sure."* Workshops, coaching sessions, online discussions, and newsletters are examples on different communication formats used [P41]. Another approach in communicating the change was that managers explained and encouraged change in an extensive series of one-on-one discussions [P22].

Some studies emphasized communicating the goals of the transformation. Having a clear message of expectations helped remove confusion and make people understand the purpose of the transformation [P43, P48]. The motivation to change can be increased by having a simple proposal on how to reach the desired outcomes [P40]. McDowell and Dourambeis [P32] describe how the organization launched a series of communication events especially designed to emphasize the reasons behind the agile practices.

Make the change transparent. Enabling transparency during the transformation was reported as important, and even highlighted as a critical factor for success [P11, P16]. Fry and Greene [P16] underline the importance of transparency and sharing of information: *"This bias to sharing information with everyone was critical in our ability to adapt on a daily basis to ensure our success."* Transparency was achieved by showing both successes and challenges [P6], actively reaching out for feedback [P6], using project management tools to display project status publicly [P11], by rearranging physical spaces [P42], and by holding the meetings of the roll-out team in public [P16]. By sharing experiences and status of the transformation the organization was aligned and everyone was moving in the same direction [P42, P43].

Create and communicate positive experiences in the beginning. Several cases highlighted that the agile transformation spread effectively through positive word-of-mouth [P6, P40]. The move towards agile is assisted by making any benefits publicly visible and celebrating even the small victories [P40, P42]. When good results were shown by a team it created interest in others, and enthusiasm to try the new way of working would spread [P39, P46]. Some companies used agile and waterfall methods side by side. This setting made comparison possible, bringing up the benefits of agile [P19, P36].

Although a recipe to fabricate positive experiences can not be given, many cases described how they had succeeded. Benefield [P6] describes how a survey was done to objectively measure satisfaction in agile, and the positive results helped teams make the decision to change their way of working. Further, management requested proof on better performance as a response to requests to increase coaching and training budget. After conducting inquiries, the coaching team was able to present an estimate of a 30% increase in performance, which made management highly convinced [P6].

Mindset and alignment

Concentrate on agile values. In a number of studies it was described that the principles of agile should be emphasized over practices and simple mechanics [P46, P48]. When people understand the agile values they will also understand why the change is being done and feel motivated [P1, P16, P42].

Some cases described that inexperienced coaches and Scrum masters made mistakes in focusing too much on the implementation of practices [P23, P30].

Arrange social events. Social events were reported to benefit the transformation by helping to build an agile mindset. A few cases even described the transformation being driven by a series of events where people received information and had the possibility to participate in shaping the new way of working [P32, P41]. One case described how the corporate agile awareness and teaching event was designed to be fun and engaging, which made people more enthusiastic in applying the agile practices [P32].

Various social activities were presented as valuable tools for improving bonding within teams [P11, P27, P40]. The importance of creating personal bonding was also highlighted for change leading entities such as management and coaching teams [P11, P23].

Cherish agile communities. The formation and influence of agile communities was reported to have a significant impact on transformation. The emergence of an agile community was reported as a success factor in the transformation [P3]. A community was also seen as indispensable as it has power to overcome impediments that would be too large for individuals to affect [P47]. The agile communities raised awareness of agile methods in the organization [P41, P42, P47], and spawned eager followers who spread the word even further [P12].

Align the organization. A factor in achieving change was creating alignment towards the common goal of introducing new development methods. It was seen as an important factor that all levels in the organization speak the same language and accept the change [P18, P31]. Alignment was built by promoting success stories and gathering lists of problems to tackle [P42, P46]. In one case, a complete alignment between teams and within management was considered as necessary in order to use agile in a large context [P43]. Another case highlighted creating and applying a structured roll-out process to uniformly introduce agile to a large number of teams [P8].

Team autonomy

Allow teams to self-organize. The agile principle of giving teams the power to decide over themselves was seen as an important factor in advancing the transformation. In some cases, management initially attempted to prescribe how the new practices should be implemented [P5, P18, P41]. However, it was learned that only when full control was given to teams the agile methods could be properly established [P5, P16, P41]. Allowing self-organization creates commitment to the change and motivation to continued use [P16, P18]. It allows teams to take ownership of the development process and voluntarily improve it even further [P41].

The acceptance of agile methods was easy when teams gained the authority to decide on development speed and quality [P45]. Favoring empowerment over prescribing the new practices was also reported to improve performance [P27]. Giving teams the authority to decide over work items increased productivity and morale [P30, P34]. Roche and Vasquez-McCall [P41] describe how prescribing the agile methods to use led only so-far in the transformation. The effectiveness of teaching and communicating the transformation started to decline. To enable the transformation to proceed further, an organization-wide challenge was created where teams would independently develop and showcase the best agile practices. As a result the ownership of the methods was transferred to teams, and the new way of working gained a secure foothold [P41].

Allow grass roots level empowerment. An interesting success factor was the absence of a top-down mandate. Atlas [P3] describes that the use of Scrum was spreading because teams were both allowed to use and enabled to learn the method. The transformation itself was agile when people felt empowerment in making the decision to adopt. The incentive to change was created when teams learned about agile methods, and perceived that a change would be beneficial [P3]. The absence of mandate granted genuine support for the new way of working on the grass roots level, which was a necessity for the process to work [P6]. It was also thought that proceeding with a top-down mandate would have made teams conform to a single process [P41]. This would have been sub-optimal, as it was important that each individual team and project tailored their practices [P41].

Requirements management

Recognize the importance of the product owner role. A particularly important role was the Product Owner. Many cases reported problems if the Product Owner did not perform adequately, and it was seen as critical to have a dedicated person in that role [P12, P14, P30, P39]. Successes or problems emerged respectively depending on how well the Product Owners were performing.

In one case, a well-implemented Product Owner role was understood to be a key success factor when using agile methods [P16, P33]. It was reported that when the Product Owner role was implemented correctly, the team performed better and the work products were correct [P24]. Projects started off better when the Product Owners were engaged early on [P46]. Some Product Owners were even described to become agile enthusiasts when they learned how business and technology can collaborate in the agile way of working [P46].

Several cases endorsed training and coaching for the Product Owner role [P16, P39, P46]. Product Owners should receive training on agile principles, backlog management, user story breakdown, and agile planning [P16]. Also tool support for backlog management and two way communication between Product Owners and teams should be worked on [P39].

Invest in learning to refine the requirements. Some cases reported challenges in requirements management, highlighting the difficulty in managing the gap between high-level requirements and user stories handled by teams. It is important that the requirements are concise and small enough for the teams to handle [P5, P11]. It was recommended to invest in teaching skills in breaking down and writing user stories [P11, P33, P48]. Gat [P17] describes how the gap between development and requirements management was bridged by having a dedicated ”requirements architect” to help with requirements refinement.

DISCUSSION

In this section we first discuss our general observations, followed by a discussion on the answers of our research questions. Then, we identify discrepancies and open issues in the literature. Finally, we discuss the limitations and present a future research agenda.

General observations

In the past decade several publications have discussed the use of large-scale agile. We identified 52 papers presenting 42 industry cases describing large-scale agile transformations. The identified primary studies were almost exclusively experience reports. We could only find six research papers that we classified as case studies. None of these papers had their research focus directly on the transformation process, even though they provided some information on the transformation process as well. As can be seen from our relevance classification in Table 12. four research papers received classification 2, while one received 3 and one 1 (with scale 1–5, where 5 means that the entire paper focuses on describing the transformation process). When comparing to experience reports, where 20 of 46 papers received 4 or 5 in our classification, we can conclude that experience reports concentrated better on our research topic, lager-scale agile transformations, than the research papers. Moreover, as the second most often mentioned challenge in our primary studies rose the "Lack of guidance from literature" which was mentioned by nine cases.

Thus, the main finding of this study is that despite the relevance of the topic for practitioners, research is seriously lagging behind, and there is in particular a need for rigorous case studies and summarizing research.

Answers to research questions

This section summarizes and discusses the answers to the research questions.

As an answer to RQ1: "What challenges have been reported for large-scale agile transformations?", we identified 35 challenges grouped into nine categories, summarized in Table 11. The challenge categories that received the most mentions were: 1) Agile difficult to implement (mentioned by 48% of the cases), 2) Integrating non-development functions (43%), 3–4) Change resistance (38%) and Requirements engineering challenges (38%). From individual challenges the most mentions received: Other functions unwilling to change (31%) and Lack of guidance from literature (21%).

These results show that large-scale agile seems to be harder to implement than people expect, as companies complain about not finding enough guidance in the literature. With increasing organization size organizational functions beyond development get involved, and they need to interface with development. Such functions range from marketing and sales to user experience and human resources. If these functions are not aligned with the transformation, that might cause serious limitations for the agile implementation and thus the full potential of the agile cannot be realized.

To answer RQ2: "What success factors have been reported for large-scale agile transformations?" we identified 29 success factors grouped into eleven categories, as presented in Table 12. The success factor categories that stand out are: 1) Choosing and customizing the agile approach (50%), 2–3) Management support (40%) and Mindset and Alignment (40%), and 4) Training and coaching (38%). From individual success factors the most often mentioned were: Ensure management support (29%), Coach teams as they learn by doing (29%),

Customize the agile approach carefully (26%) and Start with a pilot to gain acceptance. These success factors show that large-scale agile cannot be just taken into use off-the-shelf, but careful customization is needed to make a good fit for the organization and to gain the best benefits. Even though agile is often seen as a grass-root movement starting from development, management support is clearly needed when performing a large-scale agile transformation. Mindset and alignment rose high as well, emphasizing the fact that understanding the agile values behind the agile practices is important and aligning the whole organization towards the common goals makes it easier to pull through the transformation.

There does not exist similar literature studies on large-scale agile as ours, nor does there exist any surveys specifically on large-scale agile. The studies and surveys that do exist have studied agile in general, not specifically as large-scale nor agile transformations, e.g. Chow and Cao (2008) studied success factors in agile software projects in general. The only studies that touch the topic of this paper are not scientific but done by agile consulting or tool companies, e.g. Version One's State of Agile surveys (VersionOne, Inc, 2016) or Forrester's surveys (For, 2012; Giudice et al., 2014).

The closest to our study is the State of Agile survey, as large part of the respondents of their latest survey (VersionOne, Inc, 2016) were from large organizations that had at least partially adopted agile. The results of that survey have similarities with ours. Their respondents reported as the biggest barriers for further agile adoption the following: the ability to change the organizational culture (55%), general organizational resistance to change (42%), pre-existing rigid/waterfall framework (40%), not enough personnel with the necessary agile experience (39%) and management support (38%). Change resistance rose as a big challenge in our study as well, and we recognized management support as one of the top success factors. Pre-existing rigid waterfall frameworks did not rise as a separate challenge in our study, but we recognized "management in waterfall" as a challenge as well as "using old and new approaches side by side", where the old approach was most often just waterfall or waterfall type of rigid method. The ability to change the organizational culture was not recognized by our study as such, however, challenges close to that were, e.g. "management unwilling to change", "keeping the old bureaucracy" and "other functions unwilling to change". However, as a success factor category we had "Mindset and alignment", which included success factors such as "concentrate on agile values" and "align the organization". Thus, our study recognized some factors that relate to the challenges and the importance of changing the organizational culture. Finally, the barrier "not enough personnel with the necessary agile experience" was not recognized as a separate challenge in our study, but "training and coaching" was among our most important success factor categories.

The State of Agile survey did not bring up our two biggest challenge categories "Agile difficult to implement" and "Integrating non-development functions". One explanation to that might be that the survey most probably had pre-defined answer categories limiting answers to those, whereas our categories rose from the data, and thus were not limited to any pre-defined options.

The State of Agile survey gave five tips for success with scaling agile: consistent process and practices (43%), implementation of a common tool across teams (40%), agile consultants or trainers (40%), executive sponsorship (37%) and internal agile support team (35%). Four of these tips match very well to our findings. The first one of these, the consistent process and practices, is very close to our success factor "confirm to a single approach". The third one, agile consultants or trainers, match to our training and coaching category, where we have as success factors "provide training on agile methods" and "coach teams as they learn by doing". The fourth tip, executive sponsorship, fit to our success factor category "management support". Finally, the last tip, internal agile support team, is at least partially included in our success factor "coach teams as they learn by doing", as in many of the cases that mentioned this, the day-to-day coaching was provided by internal coaches. However, the second tip, implementation of a common tool across teams, did not rise as a success factor in our study, even though several cases mentioned the implementation of common tools. Nevertheless, they did not lift that up as a success factor for transformation.

Discrepancies and open issues

The papers included in our review provided several pieces of good advice in the form of challenges and success factors. However, looking at the challenges and success factors, we can notice discrepancies, and even controversial advice, highlighting aspects of agile transformation that would require deeper study to gain more insight.

Dealing with the old process. It is unclear how and when to completely drop the old process and its related concepts and practices. The literature mentions challenges when keeping the old way of working, such as using the "Old and new models side by side", "Management in waterfall mode", and "Integrating non-development functions in the transformation" indicating that the coexistence of two different ways of working is problematic. On the other hand, most transformations were done using a stepwise approach, meaning that this situation will exist at least for some while in the organization, and using the old way of working as a reference for the transformation can also be beneficial, as the good practice "Map to old way of working to ease adaptation" suggests.

Drawing the line between "one single approach" and allowing self-organization. On the one hand, to successfully perform an agile transformation, it seems important to use a single approach as a starting point, as the success factor: "Conform to a single approach" indicates. This allows the organization to have a consistent vocabulary, which was seen to benefit the organization and support the change [P8, P16, P43]. A single approach makes it possible to compare work between teams and makes it easier for people to relocate and have predictable progress [P43]. If such an approach is lacking, challenges like "interpretation of agile differs between teams", becomes exacerbated, making it difficult to collaborate between teams or change team members between the teams. On the other hand, some primary sources suggest that teams should be allowed to self-organize and customize their practices individually to fit the needs of each team, and full control should be given to teams. It was also mentioned that in a larger organization it is not feasible to use the same process for all projects [P49], but individual application of the agile process is needed for different types of development, depending, e.g., on the type of the software being developed or the project size [P45].

Providing management support without suppressing grass-root level empowerment. On the one hand, having management support is a necessary condition for successful transformation, as shown by the related success factors. However, forcing the transformation from the top can create problems, as some organizations reported "Top-down mandate creates resistance" as a challenge. Of the reported transformations, about half were led top-down, and about one forth bottom-up, and one fourth using a hybrid model.

Limitations

Researcher bias might have influenced the selection of primary studies, as well as data extraction. The selection of the primary studies may have been distorted by interpreting the inclusion criteria falsely. This risk was mitigated by using three researchers in designing the inclusion criteria. When the inclusion criteria was subsequently applied, the abstract filtering was performed independently by two researchers, and unclear cases were resolved by case-by-case discussions between two or three researchers. In the full-text filtering phase, one researcher did the initial filtering making the decision regarding the unambiguous cases. For the less clear-cut cases—over half of the papers—two researchers read the paper and independently made decision proposals regarding inclusion or exclusion. In cases they agreed, the paper was included or excluded based on their joint agreement. Papers that still were unclear were discussed and resolved by all three researchers together.

The second part of the research that may have been affected by subjective bias was the elicitation of results by coding and analysis. Our tools for mitigating this threat were limited, as the work stages in question are particularly laborious. For resource constraints the process could not be duplicated. We made an attempt to prevent subjective bias in analysis by making the results as traceable as possible, by supplying references to each claim presented in the results.

A particular problem in this systematic literature review was the limitations of Boolean keyword searches in online databases. A keyword search cannot easily identify the facets *large-scale* and *empirical*, and also the facet *transformation* is difficult to capture with keywords. For these reasons we did not include the facets *large-scale*and *empirical* in the keyword search, but instead did a manual walkthrough of all the selected papers in the filtering phase to determine the size of the organization and whether the paper contained empirical material. This added manual work, but mitigated the threat.

For the facet *transformation*, we used a variety of synonyms as keywords, but a small risk remains that some studies discussing transformations without using any of the keywords we used remain unidentified.

Through our preliminary search we found three relevant papers that were not found by our keyword search, and that we included in the study as primary sources. In addition, by going through the references of the 170 papers selected for full text filtering we identified two additional relevant papers that had not been spotted by the keyword search and that we included as primary studies. By these means we aimed to make sure to locate papers that the keyword search might have missed. The fact that it was possible to find additional papers outside the search result suggests that a possibility remains that our keyword search missed some other papers as well. However, because we put reasonable effort in attempting to identify missed publications, we believe that the number of possibly missed papers remains very limited, and thus does not significantly affect the results.

Only six of the primary studies presented a clearly defined research method. Most of the papers were experience reports, in most of which author was a member of the organization discussed, as shown in Table 6, creating author bias. However, due to the low number of research papers, we deemed that the results would be distorted heavily and many valuable studies would be left out if a strict quality assessment would be part of the inclusion criteria. As a result we decided to include all experience reports, regardless of the perceived objectivity.

Many primary studies were openly pro-agile, without giving a solid motivation for the standpoint. The tendency to publish only positive results is another particular problem for this literature review. The primary studies typically reported the transformations as successful, a sign of publication bias. A related issue was that the majority of experience reports were authored by persons personally involved in the transformation, due to which the authors might be reluctant towards reporting problems in the transformation. However, most papers did bring up several challenges experienced during the transformation as well as perceived success factors. Due to the current state of research it is necessary to include studies with varying strength of evidence in order to relevantly aggregate evidence on large-scale transformations. Due to the author and publication bias of the primary studies and qualitative nature of transformation descriptions, we decided not to make quantitative interpretations in the results, instead use only qualitative analysis. Therefore, based on this data we could not, e.g., rank the challenges and success factors and say which ones are the most important ones. However, by looking at the number of cases that mentioned some success factor or challenge gives an indication about the prevalence of the factors in the existing literature. However, as we do not know how well the published papers represent the industry, we cannot generalize e.g. the percentages to the industry at large.

Future research agenda

Based on our literature review, we identify five topics that we think requires significant further study, and that can together be considered to form a research agenda on large-scale agile transformations.

1) Case studies on transformations

More research conducted with proper methods on large-scale agile transformations is acutely needed. In our review, we were able to identify only six research papers on the topic, despite the large practitioner interest in the topic. The timeliness and importance to practitioners is evidenced, e.g., through the fact that recent years have seen a number of books published and courses organized by consultants, and a large number of talks on this topic at agile conferences. Thus, the found 46 experience reports and six research papers seem to only scratch the surface and report experiences from only a small fraction of the real number of transformations happening. Thus, we suggest more case studies on understanding the large-scale agile transformations better and how they are done in practice.

2) Scaling practices

In our review, we scoped out the practical adaptations and augmentations, "scaling practices", used by the organizations. Some agile methods, e.g., Scrum have suggestions: Scrum proposes the use of Scrum-of-Scrums as the main scaling practice. Several frameworks by consultants have been developed and are actively promoted, e.g. SAFe and LeSS. These frameworks introduce several additional practices for scaling. E.g., LeSS suggests practices like Area Product Owners and Communities of Practice. Yet, little research has been done on what scaling practices actually are used in companies, or of their related challenges and benefits or context-dependency.

3) Scaling frameworks

Consultants and practitioners have put forward several frameworks for scaling agile. For example, agile consultants have put together an "Agile Scaling Knowledgebase Decision Matrix" (Mat, 2016), where they briefly compare different agile scaling approaches in one big excel sheet. Currently, they have listed nine different approaches. While going through the articles for this review, we noticed that experiences reports and research papers very seldom mention any frameworks. To our knowledge, there exists only a handful of papers on actual usage experiences related to a particular scaling framework. This is in stark contrast to the fact that companies report using these frameworks, e.g., according to Version One's 10th survey on the state of agile 23% of respondents reported using SAFe and 4% LeSS. Thus, it would be important to study the scaling frameworks scientifically: How much are these scaling frameworks used? How are they used? What are the benefits and challenges of using them? To what kind of circumstances is each of them suitable? How much are they tailored in practice to the needs of the customers? As the reported usages of the scaling frameworks in the scientific literature is low, we encourage researchers to perform in-depth case studies to understand how the frameworks are used and how they are possibly customized.

4) Enterprise agile

As one of our challenge groups, "Integrating non-development functions in the transformation", showed the agile adoption often started from the development organization, and other parts of the organization found it difficult to adapt. However, the development organization operates in a larger context and needs to interface with other organizational functions. This is challenging if the other functions do not adapt to or adopt agile, and are unwilling to change. Thus, for ensuring the success of the whole transformation, it seems to be important that other organizational functions support and adopt agile. It would be interesting to study how these other functions can best be included in, and support an agile transformation at the enterprise scale.

5) Surveys on challenges and success factors

Surveys on challenges and success factors for agile projects in general have been conducted, e.g.(Chow and Cao, 2008). However, specifically large-scale agile projects have not been scientifically studied. Non-scientific surveys exist, the most famous one being the State of Agile Survey that Version One has been conducting annually since 2007. That survey has recently asked a few questions related to large scale as well, e.g. on scaling methods used and tips for success with scaling agile. According to the latest survey (VersionOne, Inc, 2016), 62% of the almost 4000 respondents had more than hundred people in their software organization and 43% of all the respondents worked in development organizations where more than half of the teams were agile. Of course, the sample of this study is limited to a selected subset of companies and countries (of the almost 4000 respondents to the latest survey 65% were from North America and 26% from Europe). However, this indicates that there seems to exist a large number of companies that have taken or are taking large-scale agile into use. Thus, there is a possibility to perform interesting survey studies as well. One topic for further surveys would be to study how the challenges and success factors recognized in this study are experienced in the companies: which ones they have experienced and which ones they consider most important.

CONCLUSIONS

We presented a systematic literature review of empirical studies and experience reports on large-scale agile transformations. We analyzed 52 papers describing 42 different organizations, presenting qualitative findings describing reported challenges and success factors for large-scale agile transformations.

The identified primary studies were almost exclusively experience reports, only six research papers were included. Thus, the main finding of this study is that despite the relevance of the topic for practitioners, research is seriously lagging behind, and as a consequence the identified success factors and challenges are those that practitioners perceive and report as most important. The relationship between these and objective fact remains unknown. The challenge categories that received the most mentions are agile difficult to implement, integrating non-development functions, change resistance, and requirements engineering challenges. The success factor categories that received the most mentions are choosing and customizing the agile approach, management support, mindset and alignment, and training and coaching.

As future research topics we suggest case studies on agile transformations, studies on the usage of scaling practices and scaling frameworks, as well as enterprise wide use of agile, and surveys on large-scale agile.

ACKNOWLEDGMENT

This study was funded by TEKES as part of the Need for Speed (N4S) SHOK program.

REFERENCES

1. Anonymous, Agile Software Development and the Factors that Drive Success, Tech- nical Report, Forrester Research Inc., 2012.
2. Beck, K. , 1999. Embracing change with extreme programming. Computer 32 (10), 70–77 .
3. Berger, H. , Beynon-Davies, P. , 2009. The utility of rapid application development in large-scale, complex projects. Inf. Syst. J. 19 (6), 549–570 .
4. Bjarnason, E. , Wnuk, K. , Regnell, B. , 2011. A case study on benefits and side-effects of agile practices in large-scale requirements engineering. In: Proceedings of the 1st Workshop on Agile Requirements Engineering. In: AREW '11, pp. 3:1–3:5 .
5. Boehm, B. , 2002. Get ready for agile methods, with care. Computer 35 (1), 64–69 .
6. Boehm, B. , Turner, R. , 2005. Management challenges to implementing agile pro- cesses in traditional development organizations. Softw. IEEE 22 (5), 30–39 .
7. Chow, T., Cao, D.-B., 2008. A survey study of critical success factors in agile software projects. J. Syst. Softw. 81 (6), 961–971. doi: 10.1016/j.jss.2007.08.020 .
8. Cloke, G. , 2007. Get your agile freak on! agile adoption at yahoo!

music. In: Agile Conference (AGILE), 2007, pp. 240–248 .

9. Cockburn, A., Highsmith, J., 2001. Agile software development, the people factor. Computer 34 (11), 131–133. doi: 10.1109/2.963450 .
10. Cohn, M. , Ford, D. , 2003. Introducing an agile process to an organization [software development]. Computer 36 (6), 74–78 .
11. Cruzes, D. , Dyba, T. , 2011. Recommended steps for thematic synthesis in software engineering. In: Proceedings of the 2011 International Symposium on Empirical Software Engineering and Measurement. In: ESEM '11. IEEE Computer Society, Washington, DC, USA, pp. 275–284 .
12. Dingsøyr, T., Fægri, T., Itkonen, J., 2014. What is large in large-scale? a tax- onomy of scale for agile software development. In: Jedlitschka, A., Ku- vaja, P., Kuhrmann, M., Männistö, T., Münch, J., Raatikainen, M. (Eds.), Product- Focused Software Process Improvement. In: Lecture Notes in Computer Sci- ence, vol. 8892. Springer International Publishing, pp. 273–276. doi: 10.1007/ 978-3- 319- 13835- 0 _ 20 .
13. Dingsøyr, T. , Moe, N. , 2013. Research challenges in large-scale agile software devel- opment. SIGSOFT Softw. Eng. Notes 38 (5), 38–39 .
14. Dingsøyr, T., Moe, N., 2014. Towards principles of large-scale agile development. In: Dingsøyr, T., Moe, N., Tonelli, R., Counsell, S., Gencel, C., Petersen, K. (Eds.), Ag- ile Methods. Large-Scale Development, Refactoring, Testing, and Estimation. In: Lecture Notes in Business Information Processing, vol. 199. Springer Interna- tional Publishing, pp. 1–8. doi: 10.1007/978- 3- 319-14358- 3 _ 1 .
15. Dybå, T. , Dingsøyr, T. , 2008. Empirical studies of agile software development: a sys- tematic review. Inf. Softw. Technol. 50 (9-10), 833–859 .
16. Dybå, T. , Dingsøyr, T. , 2009. What do we know about agile software development? Softw. IEEE 26 (5), 6–9 .
17. Elshamy, A. , Elssamadisy, A. , 2006. Divide after you conquer: an agile software de- velopment practice for large projects. In: Extreme Programming and Agile Pro- cesses in Software Engineering. In: Lecture Notes in Computer Science (includ- ing subseries Lecture

Notes in Artificial Intelligence and Lecture Notes in Bioinformatics), 4044 LNCS, pp. 164–168 .

18. Federoff, M. , Courage, C. , 2009. Successful user experience in an agile enterprise environment. In: Procceedings of HCI International 2009, San Diego, CA , USA , July 19-24, Part I. In: LNCS, 5617, pp. 233–242 .
19. Fitzgerald, B. , Hartnett, G. , Conboy, K. , 2006. Customising agile methods to software practices at intel shannon. Eur. J. Inf. Syst. 15 (2), 200–213 .
20. Fowler, M. , 20 0 0. Put your process on a diet. Softw. Dev. 8 (12), 32–36 .
21. Fowler, M., Highsmith, J., The agile manifesto (2001). http://www. agilemanifesto. org/ .
22. Freudenberg, S. , Sharp, H. , 2010. The top 10 burning research questions from prac- titioners. Softw. IEEE 27 (5), 8–9 .
23. Fry, C. , Greene, S. , 2007. Large scale agile transformation in an on-demand world. In: Agile Conference (AGILE), 2007, pp. 136–142 .
24. Fulgham, C. , Johnson, J. , Crandall, M. , Jackson, L. , Burrows, N. , 2011. The FBI gets agile. IT Prof. 13 (5), 57–59 .
25. Giudice, D.L. , Kisker, H. , Angel, N. , 2014. How Can You Scale Your Agile Adoption? Technical Report. Forrester Research Inc .
26. Hamed,A.,Abushama,H.,2013.Popularagileapproachesinsoftware development: review and analysis. In: Computing, Electrical and Electronics Engineering (IC- CEEE), 2013 International Conference on, pp. 160–166. doi: 10.1109/ICCEEE.2013.
27. 6633925 .
28. Highsmith, J., Cockburn, A. , 2001. Agile software development: the business of innovation. Computer 34 (9), 120–127 .
29. Hodgkins, P. , Hohmann, L. , 2007. Agile program management: Lessons learned from the verisign managed security services team. In: Agile Conference (AGILE), 2007, pp. 194–199 .
30. Jalali, S. , Wohlin, C. , 2012. Global software engineering and agile practices: a sys- tematic review. J. Softw. 24 (6), 643–659 .
31. Kaisti, M. , Rantala, V. , Mujunen, T. , Hyrynsalmi, S. , Konnola, K. , Makila, T. , Lehto- nen, T. , 2013. Agile methods for embedded

systems development - a literature review and a mapping study. EURASIP J. Embedded Syst. 2013 (1), 15 .

32. Kim, E. , Ryoo, S. , 2012. Agile adoption story from NHN. In: Computer Software and Applications Conference (COMPSAC), 2012 IEEE 36th Annual, pp. 476–
33. 481 .
34. Kitchenham, B.A. , 2007. Guidelines for performing Systematic Literature Reviews in Software Engineering. Technical report EBSE-2007-01. Keele University .
35. Koehnemann, H. , Coats, M. , 2009. Experiences applying agile practices to large sys- tems. In: Agile Conference, 2009. AGILE '09., pp. 295–300 .
36. Lindvall, M. , Muthig, D. , Dagnino, A. , Wallin, C. , Stupperich, M. , Kiefer, D. , May, J. , Kahkonen, T. , 2004. Agile software development in large organizations. Com- puter 37 (12), 26–34 .
37. Livermore, J.A. , 2008. Factors that significantly impact the implementation of an ag- ile software development methodology. J. Softw. 3 (4), 31–36 .
38. Lyon, R. , Evans, M. , 2008. Scaling up – pushing scrum out of its comfort zone. In: Agile, 2008. AGILE '08. Conference, pp. 395–400 .
39. Maranzato, R. , Neubert, M. , Herculano, P. , 2012. Scaling scrum step by step: the mega framework. In: Agile Conference (AGILE), 2012, pp. 79–85 .
40. Miller, J. , Haddad, H. , 2012. Challenges faced while simultaneously implementing cmmi and scrum: a case study in the tax preparation software industry. In: In- formation Technology: New Generations (ITNG), 2012 Ninth International Con- ference on, pp. 314–318 .
41. Mishra, D. , Mishra, A. , 2011. Complex software project development: agile methods adoption. J. Softw. Maint. Evolution 23 (8), 549–564 .
42. Misra, S.C. , Kumar, V. , Kumar, U. , 2010. Identifying some critical changes required in adopting agile practices in traditional software development projects. Int. J. Qual. Reliab. Manag. 27 (4), 451–474 .
43. Moore, E. , Spens, J. , 2008. Scaling agile: Finding your agile tribe.

In: Agile, 2008. AGILE '08. Conference, pp. 121–124 .

44. Nerur, S. , Mahapatra, R. , Mangalaraj, G. , 2005. Challenges of migrating to agile methodologies. Commun. ACM 48 (5), 72–78 .
45. Paasivaara, M., Behm, B., Lassenius, C., Hallikainen, M., 2014. Towards rapid re- leases in large-scale xaas development at ericsson: A case study. In: Global Soft- ware Engineering (ICGSE), 2014 IEEE 9th International Conference on, pp. 16–25. doi: 10.1109/ ICGSE.2014.22 .
46. Paasivaara, M. , Durasiewicz, S. , Lassenius, C. , 2008. Using scrum in a globally dis- tributed project: a case study. Softw. Process Improv. Pract. 13 (6), 527–544 .
47. Paasivaara, M., Lassenius, C., Heikkila, V., Dikert, K., Engblom, C., 2013. Integrating global sites into the lean and agile transformation at ericsson. In: Global Soft- ware Engineering (ICGSE), 2013 IEEE 8th International Conference on, pp. 134 – 143. doi: 10.1109/ ICGSE.2013.25 .
48. Petersen, K. , Wohlin, C. , 2010. The effect of moving from a plan-driven to an incre- mental software development approach with agile practices. Emp. Softw. Eng. 15 (6), 654–693 .
49. Sagesser, K. , Joseph, B. , Grau, R. , 2013. Introducing an iterative life-cycle model at credit suisse it switzerland. Softw. IEEE 30 (2), 68–73 .
50. Schwaber, K. , Beedle, M. , 2002. Agile Software Development with Scrum. Series in Agile Software Development. Prentice Hall .
51. Scott, J. , Johnson, R. , McCullough, M. , 2008. Executing agile in a structured or- ganization: government. In: Agile, 2008. AGILE '08. Conference, pp. 166–170 .
52. Senapathi, M. , Srinivasan, A. , 2013. Sustained agile usage: a systematic litera- ture review. In: Proceedings of the 17th International Conference on Evalu- ation and Assessment in Software Engineering. In: EASE '13. ACM, pp. 119–124 .
53. Spearman, S. and Dolman, R. Agile scaling knowledgebase decision matrix, 2016.
54. http://www.agilescaling.org/ask-matrix.html. Cited 10.6.2016.
55. VersionOne, Inc, 10th annual"state of agile development" survey,

2016.

56. Williams, L. , Cockburn, A. , 2003. Agile software development: it's about feedback and change. Computer 36 (6), 39–43

REFERENCES

1. B. Fitzgerald and K. J. & Stol, "Continuous software engineering and beyond: trends and challenges," in Proceedings of the 1st International Workshop on Rapid Continuous Software Engineering, 2014, pp. 1 -9.
2. L. Chen, "Continuous delivery: Huge benefits, but challenges too," Software, IEEE, vol. 32, no. 2, pp. 50-54, 2015.
3. G. G. Claps, R. B. Svensson, and A. Aurum, "On the journey to continuous deployment: Technical and social challenges along the way." Information and Software technology, vol. 57, pp. 21 -31, 2015.
4. H. H. Olsson and J. Bosch, "Towards agile and beyond: an empirical account on the challenges involved when advancing software development practices," in Agile Processes in Software Engineering and Extreme Programming. Roma, Italia: Springer International Publishing, 2014, pp. 327-335.
5. E. Dustin, T. Garrett, and B. Gauf, Implementing

automated software testing: How to save time and lower costs while raising quality: Pearson Education, 2009.

6. S. Elbaum, G. Rothermel, and J. Penix, "Techniques for improving regression testing in continuous integration development environments," in Proceedings of the 22nd ACM SIGSOFT International Symposium on Foundations of Software Engineering, Hong Kong, China, 2014, pp. 235-245.
7. M. Cohn, Succeeding with agile: software development using Scrum. Pearson Education, 2010. A. Bagmar, "Enabling Continuous Delivery in Enterprises with Testing," in Agile2015, India, 2015.
8. Armbrust M, et al. Above the Clouds: A Berkeley View of Cloud. California, USA: University of California at Berkley, 2009. http://www.cs.columbia.edu/~roxana/teaching/COMS-E6998-7-Fall-2011/papers/armbrusttr09.pdf
9. J. Liberty and D. Maharry, Programming ASP.NET 3.5, 4th ed. Massachusetts, USA: O'Reilly Media, 2008.
10. Mell P; Grance T. The NIST Definition of Cloud Computing. Gaithersburg, USA: Rebecca M, 2011.
11. Rountree D; Castrillo RESEARCHER. The Basics of Cloud Computing: Understanding the fundamentals of Cloud Computing. Massachussets (USA): Syngress (2013).
12. Tyler C, et al. Jenkins CI: An extendable open source continuous integration server Boston: Drupal, 2014.
13. Johnson,K. (Re-)Introducing the Puppetlabs-Registry Module [blog]. San Francisco (CA): Puppet Enterprise, 19 03 2013. https://puppetlabs.com/blog/howmaintain-your-puppet-enterprise-console
14. A. M. Memon, RESEARCHER. Banerjee, and A. Nagarajan, "GUI ripping: Reverse engineering of graphical user interfaces for testing," in Proceedings of The 10th Working, Florida, USA, Nov, 2003.
15. Q. Xie and A. M. Memon, "Designing and comparing automated test oracles for GUIbased software applications," ACM Transactions on Software Engineering and Methodology, vol. 16, no. 1, p. 4, Febrero 2007.
16. S. R. Choudhary, Versee H., and A. Orso, "WEBDIFF: Automated

Identification of Crossbrowser Issues in Web Applications," in Software Maintenance (ICSM), Timişoara, Romania, 2010.

17. A. Hori, S. Takada, H. Tanno, and M. Oinuma, «An Oracle based on Image Comparison for Regression Testing of Web Applications,» in International Conference on Software Engineering and Knowledge Engineering (SEKE), Pittsburgh, USA, 2015.
18. E Selay, Z. Q. Zhou, and J Zou, «Adaptive random testing for image comparison in regression web testing,» in Digital lmage Computing: Techniques and Applications (DICTA). International Conference. IEEE, 2014, pp. 1 -7.
19. T. Saar, M. Dumas, M. Kaljuve, and N. Semenenko, Browserbite: cross-browser testing via image processing, Abril 2015, Software: Practice and Experience.
20. S. Mahajan and W. G. Halfond, "Finding HTML presentation failures using image comparison techniques," in Proceedings of the 29th ACM/IEEE international conference on Automated software engineering, 2014, pp. 91 -
21. Scrum. Professional Scrum Master RESEARCHER. Scrum.org. https://www.scrum.org/Courses/Professional-Scrum-Master
22. Beck, K; et al. Principles behind the Agile Manifesto. USA: Ward Cunningham, 2001.
23. Ferguson J. "Jenkins: The Definitive Guide." 1a edition. USA. O'Reilly Media, 2011.
24. Humble J; Farley D. "Continuous Delivery: Reliable Software Releases through Build, Test, and Deployment Automation." 1a edition. Massachusetts (USA): Pearson education Inc., 2010.
25. Kim G; Behr K; Spafford G. "The Phoenix Project: A Novel about IT, DevOps, and helping Your Business Win". 1a edition. USA. It Revolution Press, 2013.
26. Khosrow-Pour M. "Systems and Software Development, Modeling, and Analysis: New Perspectives and Methodologies". 1a edition. USA. Igi Global, 2014.
27. Schwaber K; Sutherland J. Scrum Guide, Massachusetts (USA): Scrum.org, 2013. https://www.scrum.org/Scrum-Guide
28. Turnbull J; McCunePro J; "Puppet (Expert's Voice in Open Source)"

1a edition. New York (USA), 2011.

29. Chrissis, Mary Beth. Konrad, Mike. Shrum, Sandy. CMMI®: Guidelines for Process Integration and Product Improvement; Addison Wesley Professional; The SEI Series in Software Engineering. 2003. ISBN: 0321 1 54967
30. Beck, Kent. Extreme Programming Explained: Embrace Change Addison-Wesley Professional. 1 999. ISBN: 0201 61 641 6
31. Beck, Kent. Fowler, Martin. Planning Extreme Programming. Addison-Wesley Professional. 2000. ISBN: 0201 71 091 9.

CITATIONS

CHAPTER 9

Teerat Pitakrata, Dušan Okanovića, Andrévan Hoorna, LarsGrunske, Hora: Architecture-aware online failure prediction, https://doi.org/10.1016/j.jss.2017.02.041

CHAPTER 10

Lianping Chen, Continuous Delivery: Overcoming adoption challenges, Journal of Systems and Software, Volume 128, 2017, Pages 72-86, ISSN 0164-1212, http://dx.doi.org/10.1016/j.jss.2017.02.013.

CHAPTER 11

Jesse Yli-Huumo, Andrey Maglyas, Kari Smolander, How do software development teams manage technical debt? – An empirical study, Journal of Systems and Software, Volume 120, 2016, Pages 195-218, ISSN 0164-1212, http://dx.doi.org/10.1016/j.jss.2016.05.018.

CHAPTER 12

Kim Dikert, Maria Paasivaara, Casper Lassenius, Challenges and success factors for large-scale agile transformations: A systematic literature review, Journal of Systems and Software, Volume 119, 2016, Pages 87-108, ISSN 0164-1212, http://dx.doi.org/10.1016/j.jss.2016.06.013.

INDEX

A

Adopting agile software development 280
Agile methods 279, 280, 282, 283, 288, 290, 296, 300, 303, 327, 345
Agile software development 280, 281, 286, 344, 345, 346, 348
Architectural Dependency Model (ADM) 113, 117
architectural technical debt (ATD) 209
ArgoUML 81, 82

C

capability maturity model (CMM) 204, 263, 264
Challenges 280, 291, 304, 306, 312, 318, 319, 346, 347
Cloud Computing 57
code and style sheets (CSS) 26
code smells 205
collaboration 108
Communication 86, 95
Conditional Probability Table (CPT) 121
Continuous customer 96
continuous delivery 34, 35, 36, 37, 67, 68, 71
Continuous Delivery 11
Continuous Delivery (CD) 157, 158, 193
continuous integration 67, 68
Continuous Software Development (CSD) 1

D

database (DB) 117, 121
Definition of Done (DoD) 49
design document (SDD) 39
development methodology 33, 34, 36, 37, 44
Document Object Model (DOM) 26

E

Exploratory case study 202
Extreme Programming promotes 96
Extreme Programming (XP) 282

F

Failure Propagation Model (FPM) 113, 120

G

geographical dispersion 85
graphical interface (UI) 2

I

implementation 107
Infrastructure as a Service (IaaS) 59

L

Large-scale agile, 280

M

Managing technical debt (MTD) 208
metaphor 104, 105
Microsoft Office 81

O

Organizational transformation 280, 286, 288

P

PuppetDB 81

Q

QualDev software 88
Quality of service (QoS) 112

R

RESEARCHER 81

S

software architect 218, 225, 226, 228, 229, 230, 231, 234, 235, 236, 237, 238, 244, 247, 249
Software as a Service (SaaS) 59
SonarQube 211, 227, 228, 229, 230, 231, 232, 234, 235, 238, 241, 242, 243, 244, 247, 254, 255, 256, 257, 275
Source code management 68
specification of requirements (SRD) 38
Success factors 280, 291, 320
Systematic literature review 280

T

team manager 218, 223, 224, 225, 226, 227, 228, 230, 231, 233, 235, 236, 238, 239, 240, 244, 247, 249, 250, 258
Technical debt 201, 202, 203, 205, 207, 208, 247, 264,

265, 272, 273, 275, 276
Technical debt management 202, 203, 208, 264
Technical debt management (TDM) 203, 208
technical debt (TD) 205, 208
Technical debt (TD) 201, 202

U

User interface testing (UI) 26